REX BEDDIS

The
Third World
Development and interdependence

OXFORD UNIVERSITY PRESS

Acknowledgements

Bryan and Cherry Alexander, p. 33;   Allsport, p. 171 (bottom);   Art Directors' Photo Library, pp. 56, 57 (both), 61, 69, 71 (right), 103, 111 (all), 148 (both), 150 (top, both), 152, 155 (top left), 156, 157 (both), 166 (bottom), 180 (bottom), 181 (bottom), 187 (both); Associated Press, pp. 93, 207; Christopher Bain, p. 60 (bottom);   Balfour Beatty, p. 144 (both);   Rex Beddis, pp. 57 (bottom), 65 (bottom), 164, 165 (left), 169 (top);   Jane Benton, pp. 182 (both), 183 (right);   J. A. Binns, p. 24 (top);   Nick Black, p. 176;   Camerapix, p. 149; Camera Press, pp. 66 (top), 186;   Colorific!, pp. 49, 66 (bottom);   De Beers, p. 84; Mark Edwards, p. 58;   The Financial Times, p. 168;   Format Photos, pp. 7 (bottom), 15 (right), 16 (bottom left), 55 (both), 71 (left), 78, 90 (left), 105 (bottom), 107 (both), 110, 133 (bottom), 135, 136 (bottom left), 137 (bottom), 141 (both), 142, 143, 145, 172 (right), 173, 174 (both), 175 (bottom), 177 (left), 178 (bottom), 179 (both), 190 (bottom right), 191 (all), 195, 213 (all), 215 (bottom), 229;   Paul Forster, p. 60 (top);   John Freeman, p. 224 (top); Gamma, p. 42 (bottom);   Dr D Gellner, pp. 36 (top), 37; Alastair Gray, p. 10 (all); Richard and Sally Greenhill, p. 53;   John Griffiths, p. 175 (top);   Susan Griggs, pp. 28, 30, 43, 102, 127, 140, 180 (top), 181 (bottom), 197, 209 (bottom);   Chris Hamnett, p. 44; Robert Harding, pp. 63 (right), 150 (bottom);   John Hilleson p. 31;   Jimmy Holmes, p. 36 (bottom);   Libby Howells, p. 62;   Hull City Art Gallery, 82 (bottom);   Hutchinson Picture Library, pp. 24 (right), 39 (bottom), 97, 106, 124, 138, 155 (bottom), 166 (top), 167, 169 (bottom), 170, 177 (right), 209 (top);   J. J. and P. Hubley, P. 225; IDG/Aerocamera Hofmeester, p. 51 (bottom); Impact, p. 20 (left); Suresh Karadia, p. 73 (top); Link Photos, pp. 18, 88;   Mansell Collection, pp. 80, 82 (top), 224 (bottom right);   Magnum Photos, pp. 8, 14 (top), 15 (left), 19 (bottom), 40, 48, 75, 101, 104, 108, 130, 133 (top), 139 (both), 172 (left), 185 (bottom);   J. Madeley, p. 162;   Steve McKenna, p. 137;   Marcus Moenda, pp. 16 (top right), 68 (bottom);   Ian Murphy, pp. 65 (top), 86;   Network, pp. 7 (top), 16 (bottom right), 20 (right), 23, 24 (bottom), 25 (left), 26 (both), 27, 52, 59 (both), 64, 79, 153, 158 (top), 159 (top), 160 (both), 161 (both), 178 (top), 190 (top and bottom left), 198 (both), 215 (top);   New Internationalist, p. 90 (right);   Michael O'Brien, pp. 125 (top), 128, 129, 131 (top);   Oxfam, pp. 184, 185 (top);   M. and C. Pavageau, P. 125 (bottom);   Maggie Pearson, p. 339 (top);   Popperfoto, p. 105 (top);   Punch, p. 83; Rappho, p. 42 (top);   Rex Features, pp. 29 (left), 38, 51 (top), 94, 100 (bottom), 147 (right);   Don Robertson, pp. 158 (bottom), 159 (bottom), 163 (top);   Claude Sauvageot, p. 63 (left);   Save the Children Fund, p. 98;   Bernard Smith, p. 165 (left);   Spectrum, pp. 68 (top), 154;   Frank Spooner, pp. 9 (top), 19, (top), 29 (right), 45, 73 (bottom), 76 (both), 100 (top), 131 (bottom), 147 (left), 152 (bottom), 155 (top right), 171 (top), 178 (centre), 227;   Stock Photobank, p. 14 (bottom);   Charles Swithinbank, pp. 32 (left) 34, 35;   John Taylor, p. 151;   The Times, pp. 95, 163 (bottom);   Tropix, p. 46 (bottom);   Penny Tweedie, pp. 181 (top), 183 (left); Charlotte Ward-Perkins, pp. 9 (bottom), 16 (top left), 92, 132, 136 (bottom right), 190 (top right);   Susan Watts, pp. 46 (top), 47;   Zefa Photos, p. 89.

Cover photo: Apa Photo Agency, Singapore.

Introduction

It is in the nature of publishing that the Introduction to a book is often written last of all. If Rex Beddis had lived he would, I am sure, have taken advantage of the space available to justify this book's existence (and geography's place in the curriculum) in terms of his fiercely-held belief that geography should not only establish an interest in and concern for space but also address itself to the major issues facing society. Young people should then be enabled to explore their attitudes towards these issues.

Rex was always at the forefront of thinking in geographical education, and it is no coincidence that many of his prerequisites for a vital and relevant geography have found their way into GCSE syllabuses. I have no doubt that this book which is so much a personal statement will also find its way into the classroom as a practical tool.

To demonstrate how Rex Beddis anticipated the demands of the GCSE national criteria, I shall close with an extract from an article he wrote in 1982:

> 'By combining a greater sensitivity to place with a more realistic explanation of the causes of patterns and processes, an acknowledgement of the importance of the political unit, and an involvement of attitudes towards issues, geography could be both enjoyable and important for young people.'

Tony Dale
Oxford University Press

21 Brandt reporto.

Chapter 1

Contrasts in development

There are many similarities and differences in life-styles and living standards around the world. These may be within and between countries.

These pictures show poverty in London (top), and Bombay, India (bottom)

1.1 Similarities and differences

In the early 1980s there were well over 4000 million people in the world. During the decade population increase has been roughly 70 million a year, so by 1990 there are likely to be nearer 5000 million men, women and children inhabiting the earth.

To someone from another planet the most striking thing about these millions of people would probably be how similar they are when compared with other living creatures. They look more or less alike, and their life cycles follow a fairly standard pattern. After birth the majority go through periods of infancy, childhood, maturity and old age before death, although the numbers that survive infancy and the average age at death does vary considerably from place to place. It also seems that people from all over the world have similar experiences of pleasure and pain, happiness and sadness, and anxiety about or hopes for the future. Each person is unique, and there are some recognisable differences between groups of people, but it is remarkable that thousands of millions of human beings have so much in common.

From this world point of view the similarities are obvious, but in our daily lives we are more often aware of the differences between people. There are contrasts between the appearance and behaviour of the very young and the elderly, for example. Another contrast is between males and females, although there is disagreement about whether this is due to biological differences or the learned customs and habits of various societies. It is certainly true that contrasts between men and women in terms of expected patterns of work, behaviour and responsibility are more marked in some societies than in others.

Another very obvious difference is in skin colour and facial characteristics. For some people these contrasts in appearance seem to matter a lot, and they react with suspicion, fear or contempt to others who look different. On the other hand many people welcome these differences, and value people for their personal qualities rather than just their age, sex or what they happen to look like.

A street scene in Hong Kong, one of the most crowded places on earth

The wearing of traditional clothing may give Westerners the impression that women in Muslim countries live more restricted lives than they actually do

Differences in culture and belief

Individuals and groups differ in their spoken and written language, the clothes they wear, the food they eat and in many customs and habits. Traditional forms of music, dance, drama and art vary greatly around the world. People hold very different religious and political beliefs, and these affect their views of the world and hopes for the future. It is even sometimes claimed that certain groups of people are hard-working or lazy, trust-worthy or dishonest, kind or cruel, clever or stupid and so on. These are usually completely untrue or, at least, crude generalisations, with many individual exceptions. These cultural contrasts exist not only between coun-tries but also within them. Most coun-tries have easily recognisable cultural, political and religious minority groups within their boundaries.

These differences are often under-stood and accepted, but they sometimes lead to anxiety, tension, fear and vio-lence. It is understandable that some people wish to live amongst others just like themselves, but many are confident and happy living in culturally mixed groups. It would be a very strange world if everyone looked alike, behaved in the same way, thought the same thoughts and held the same beliefs.

The changing pattern of world religions

Religion	Per cent of total world population		
	1900	1980	2000 (est)
Christian	34.4	32.8	32.2
Roman Catholic	*16.8*	*18.5*	*18.7*
Protestant or Anglican	*9.4*	*7.9*	*7.0*
Eastern Orthodox	*7.5*	*2.8*	*2.4*
Others	*0.7*	*3.6*	*4.1*
Muslim	12.4	16.5	19.2
Hindu	12.5	13.3	13.7
Buddhist	7.8	6.3	5.7
Chinese Folk Religions	23.5	4.5	2.8
Tribal and Shamanist	7.3	2.4	1.8
New Religions	0.4	2.2	2.2
Jewish	0.8	0.4	0.3
Other religions	0.8	0.8	1.0
Non-religions and atheist	0.2	20.8	21.3

Most religions are found in all countries, but concentrated in certain regions of the world. In the 1980s Christianity had a non-white majority for the first time in 1200 years.

1 List some of the things, other than age or sex, that makes one person different from another.
2 a) List four of the religions named in the table, and for each give the name of a country where it is the main belief. b) Describe and locate different religious centres in your neighbourhood or town.
3 Describe a few things or customs you use and enjoy that originated in a culture outside Britain.
4 Many Islamic women choose to wear a veil and to lead a life very different from men in their country. Do you think they are wise or foolish to do so? Why is it difficult, or perhaps wrong, to make judgements about the behaviour or beliefs of others of a different sex, religion or culture?

Exercises on page 190
Photographs illustrating the theme 'similarities and differences'

This huge statue of the Buddha is in Rangoon, Burma. Buddhism is a major world religion

1.2 Describing the world: bias in pictures and words

Pictures

In most books photographs are important sources of evidence. Those on this page show how carefully they have to be interpreted. Rio de Janeiro is a very large city of over five million people – about five times the size of Birmingham – in the South American country of Brazil. It has grown dramatically in recent decades, and while the local authority has been able to build houses for some people, over two million have built their own 'squatter' homes on the hillsides around the city. These provide a type of housing quite distinct from the slums, modest private and local authority homes and luxury dwellings found in other parts of the city. These squatter settlements around Rio are known as *favelas*.

The problem is, how can an accurate and balanced picture of the *favelas* be given with only one or two pictures? Those on this page are not faked – the places actually existed when the photographs were taken. But the photographers selected some views rather than others, and the author made further selections to convey an impression of the housing – his impression. The photographs were also taken some years ago, and these views would almost certainly be different now. The photographer and the author have selected

from many possibilities, and in that sense the impression given is biased. All pictures in books and on television and in newspapers are biased in this sense.

1 What other evidence would you like to have before judging whether these pictures of *favela* housing were accurate or not? What other evidence would it be important to have to gain a more accurate picture of the *favelas* of Rio de Janeiro?
2 What is likely to influence a photographer or author to select some pictures rather than others? If page space allowed only one of these three to be used, which would you select – and why?
3 Plan or take two series of six pictures of your neighbourhood or town to show contrasting impressions of the place – 'good' and 'bad', resident and visitor, young person and retired person, for example. Describe the ways in which each series is biased. Are your selections 'wrong', or is it enough to advise viewers to treat them with caution?

These are views of the squatter settlements on the edge of Rio de Janeiro, Brazil

Words – facts or opinions?

Writers select from what they see and describe places from a personal viewpoint, just like photographers, and this shows in what they choose to write about and in the words they use. On some occasions they try to ignore their personal opinions and feelings, and describe places exactly as anyone else would, but it is very difficult to be 'objective' in this way. It would be most unusual to find two people describing a place in the same way.

These personal differences are usually greater when it comes to explaining past or present events, and such explanations depend very much on the beliefs and attitudes of the writer. It is well known that two people can describe and explain what has happened in completely different ways. The descriptions and explanations you read are biased in the sense that they are often one person's view and interpretation. Written evidence, like picture evidence, needs to be read with caution.

A left-wing poster from mid-1920s China. Chinese peasants, workers, soldiers, merchants and students are called to unite against warlords and imperialialists

The city of Shanghai

The town of Shanghai was established 800 years ago. After the Opium War of 1840, it came under the iron heel of imperialism. Because of its favourable location, the imperialists chose it as a jumping-off point for further inroads into other parts of China. It was here that they forcibly set up foreign concessions, stationed troops, seized the Chinese maritime customs, and established banks and business firms. In collusion with Chinese feudal landlords and bureaucrat-capitalists, they engaged in speculative and criminal activities, lorded it over the Chinese people and ruthlessly fleeced them. Old Shanghai was also infested with a multitude of exploiters and parasites who looked upon this crime-ridden port as an adventurers' paradise. To the labouring people, however, Shanghai was a hell on earth. Thousands were faced with insecurity and the constant threat of unemployment, and thousands had to struggle along on the verge of starvation.

4   Read the extract about Shanghai. What do the following words or phrases mean? 'came under the *iron heel* of *imperialism*'; '*in collusion with*'; '*feudal* landlords'; '*bureaucrat-capitalist*'; '*lorded it over*'; 'ruthlessly *fleeced* them'; '*infested with* a multitude of exploiters and *parasites*'. From the use of these particular words, what do you think are the writer's feelings about the events described?

5   Say which of the following sentences are 'fact', which 'opinion', and which might be either. How can you decide what is fact and what opinion?
   a) 'Shanghai was established 800 years ago.'
   b) 'It was here that they forcibly ... stationed troops ... established banks and business firms.'
   c) 'they ... lorded it over the Chinese people and ruthlessly fleeced them.'
   d) 'exploiters and parasites who looked upon this crime-ridden port as an adventurers' paradise.'

6   The writer was Chinese. Should that make the description and explanation more or less reliable? Say why. The writer was a communist. Should that make the description more or less reliable? Say why. Does the date of the writing matter? Say why.

7   Agree with several friends to describe either a scene or an event without talking about it beforehand. Compare your descriptions and explanations.

Exercises on page 191
Images of India

An extract from *A Concise Geography of China* by Jen Yu-ti, published in Peking in 1964

11

1.3 Describing the world: bias in maps and statistics

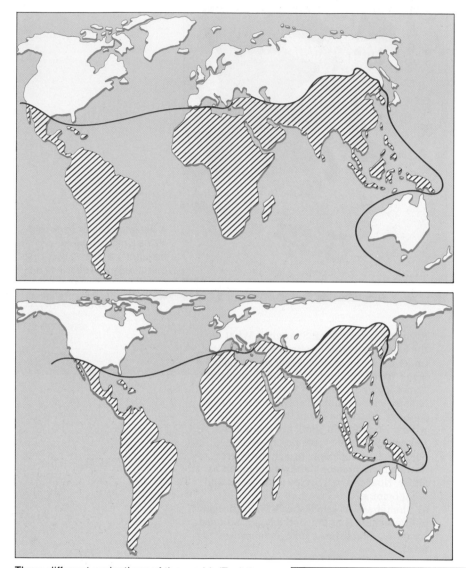

Three different projections of the world. (Top) the Mercator. (Middle) the Peters. The shaded area on these maps shows the Third World. (Bottom) a world map centred on the Pacific Ocean

Maps

It is impossible to show the world on a map without some distortion. The problem is that the earth is a sphere while a map is drawn on a flat surface. The only way to get a fairly accurate idea of the relative positions and shapes and areas of places is on a globe.

Different methods are used to try and solve the impossible problem. The mathematics used is usually very complicated, but a general idea of how maps are made can be got by imagining the globe is illuminated from within so that shapes on the surface are 'projected' onto a flat sheet of paper which is wrapped around it – the different projections are due to the source of light and the surrounding paper being in varied positions. The important point is that some world maps are suitable for some purposes and not for others. The Mercator projection is excellent for navigation since the directions or bearings from one place to another are accurate, but it greatly distorts the relative areas of places. The Peters projection shows areas in correct proportions, but badly distorts shapes. All maps suffer from these distortions, but they are most marked on maps of the whole world. It is important to realise the strengths and weaknesses of world maps before accepting the information on them.

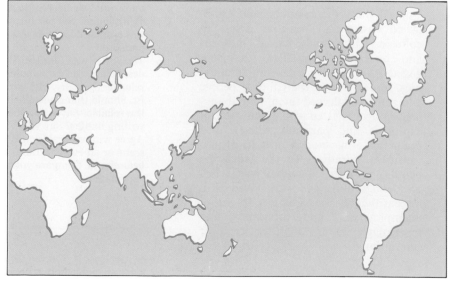

The evidence shown on the maps is also selective, of course, and is only as reliable as the accuracy of the data, its date, and what the map-maker has decided to put in or leave out. Maps are as biased as pictures and words.

Statistics

A great deal of information we receive about the world is in the form of statistics or diagrams based on statistics. As with pictures, words and maps, statistics have to be treated with caution. For example, they may have been collected in an unreliable way or they may be dated. Even when the figures are reliable, accurate and up to date, selections usually have to be made from the mass of available data. Very often those that prove a point are chosen, while others that might make a view less convincing are ignored.

There may be bias not only in the selecting of statistical data, but also in the way in which it is presented. It is well known that all sorts of people cleverly use statistical diagrams, graphs and maps to give a particular impression they favour. Statistics, whether in the form of numbers or graphs, diagrams or maps, must be read with great care.

1 Compare the representation of Europe and Africa on the Mercator and the Peters maps. What is the different visual impression given by the maps? Why do you think some people argue that it is important to use 'equal area' maps, with correct relative sizes of countries, even if the shapes are untrue?
2 Name a few countries that are likely to favour the world map centred over the South-West Pacific? Why do you think they would?
3 Why do we rarely see world maps centred over the North or the South Poles?
4 What information on the statistical map 'A World apart' gives a clue about its reliability? How can you judge whether a source is reliable or not?
5 Is any important information missing – what else would you like to know about the statistics before thinking about them?
6 a) Check on the meanings of GNP per capita, infant mortality per 1000 and the Third World from later sections of this book. b) Do the statistics support the claim made on the map?
7 Even if the statistics are reliable, accurate and up to date, why should they be treated with caution when comparing the four named areas of the world?

> **Exercises on page 192**
> Population maps of West Africa and Nigeria

These statistics reveal the enormous differences in standards of living between the continents

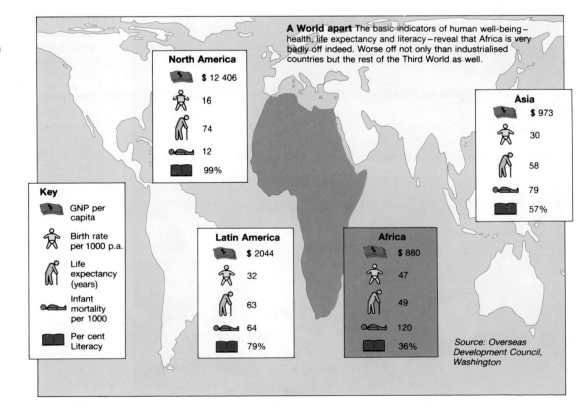

A World apart The basic indicators of human well-being – health, life expectancy and literacy – reveal that Africa is very badly off indeed. Worse off not only than industrialised countries but the rest of the Third World as well.

North America
- GNP per capita: $ 12 406
- Birth rate per 1000 p.a.: 16
- Life expectancy (years): 74
- Infant mortality per 1000: 12
- Per cent Literacy: 99%

Asia
- GNP per capita: $ 973
- Birth rate per 1000 p.a.: 30
- Life expectancy (years): 58
- Infant mortality per 1000: 79
- Per cent Literacy: 57%

Latin America
- GNP per capita: $ 2044
- Birth rate per 1000 p.a.: 32
- Life expectancy (years): 63
- Infant mortality per 1000: 64
- Per cent Literacy: 79%

Africa
- GNP per capita: $ 880
- Birth rate per 1000 p.a.: 47
- Life expectancy (years): 49
- Infant mortality per 1000: 120
- Per cent Literacy: 36%

Key
- GNP per capita
- Birth rate per 1000 p.a.
- Life expectancy (years)
- Infant mortality per 1000
- Per cent Literacy

Source: Overseas Development Council, Washington

1.4 Work and wealth

Differences in work and income

One of the great differences between people is the work they do and the effort and skill that goes into it. Many men and women have to spend hours each day in back-breaking toil on farms and plantations, or in mines, quarries or factories. Elsewhere the same jobs may be made easier by the use of machines and new technology, or less exhausting because of more humane conditions of employment. Some people choose to work at physically exhausting jobs, or in unpleasant or dangerous conditions, or for very long hours for the extra income it brings, whereas others undergo long periods of training or accept heavy responsibilities for a mixture of job satisfaction and high income.

Millions of men, women and children endure their daily drudgery to keep themselves and their dependents at little more than survival level, whilst the unemployed in many parts of the world are even worse off, and survive on the charity of others. However, a large proportion of the world's population earns more than is required for basic needs and can afford a few luxuries, and a relatively few people, either through their talent, hard work or good luck have considerable wealth. In all parts of the world there are some who are able to live on money they have inherited or invested, while others have acquired their wealth from the hard work and exploitation of others.

There are great differences not only in the work people do but also in the rewards they get for their efforts, skills and responsibilities. These differences are reflected in things such as the amount and quality of food available, in education and health care, and in the sort of housing that can be rented or bought.

Differences in Britain. Unemployed workers scavenging for a living on a rubbish tip (top). Workers at the Stock Exchange; where some of the wealthiest people work (bottom)

1   a) Give three examples of different people doing similar jobs but getting unequal rewards for it. b) Name one job that is well paid because it demands very high skills and long training; one because it involves dangerous or unpleasant conditions of work; and one because it involves responsibility for many other people.
2   Put the following in rank order, with the reason most deserving of a high income at the top and least deserving at the bottom: works long hours; uses lots of physical energy; is very skillful; is responsible for a large family; has been employed a long time; involves many years of training; is doing a socially useful job; saves lots of public money; is a good manager of others; other named reasons.

Differences in India. Skilled engineers in an aerospace factory (above left), and women labourers on a building site in Bombay (above)

Comparing the wealth of countries

It is very obvious that the wealth of people within all countries varies a great deal, although in some it is much more marked than others. Great wealth and poverty can exist side by side in most parts of the world. But the work done by *all* the people in a country can be considered to produce the wealth of that country, and this clearly varies from country to country and around the world. In general terms all countries can be considered wealthy or poor relative to others.

Quite apart from the difficulty of getting reliable information, however, it isn't easy to compare the wealth of countries with any accuracy or real meaning. One method often used is to calculate the total wealth produced within every country during a year from the goods it has made and the services provided, producing a figure known as the Gross National Product (GNP). The GNPs of countries can then be used to get some idea of their comparative wealth.

Just as important as the total wealth, though, is how this relates to the number of people in the country. If the GNP is divided by the number of people in the

country, the GNP per capita, or person, is obtained. This gives another and more useful measure of wealth for comparison. In practice, of course, the GNP is not evenly shared out, and GNP per capita masks very great differences of wealth within a country. Nevertheless, used with proper caution, GNP per capita gives a rough idea of the comparative economic wealth of countries.

3 Rank the countries in the table according to **a)** GNP, and **b)** GNP per capita in 1985.

Exercises on page 193
The world according to GNP

The wealth of ten countries compared

Country	GNP (millions US$) 1985	Population (millions) 1985	GNP per capita (US$) 1985
Australia	171 114	15.8	10 830
Bhutan	192	1.2	160
Colombia	37 488	28.4	1 320
Ethiopia	4 653	42.3	110
India	206 575	765.1	270
Mexico	163 904	78.8	2 080
Papua New Guinea	2 380	3.5	680
Spain	165 594	38.6	4 290
United Kingdom	477 990	56.5	8 460
USA	3 993 917	239.3	16 690

1.5 Living standards

The different wealth of people shows in their ability to afford adequate food and water, education and health care and proper housing – the basic needs of all people. The proportion of people who can be described as having access to these varies greatly from country to country.

Housing

The quality of housing can vary enormously even in a relatively prosperous country such as Australia, as these photographs show. They are by no means the poorest or the most expensive of housing in that country. Information from the United Nations for the 1987 International Year of Shelter for the Homeless suggests that some 100 million people in the world have no shelter of any kind! In the prosperous United Kingdom 78 000 families are officially classified as homeless, while four million households live in sub-standard or inadequate housing. In the Third World it is claimed that over half the population of the cities live in appalling slums or in squatter settlements.

Disease is an almost inevitable result of overcrowded, poorly serviced and badly located housing. This is particularly so in countries where clean running water is a rarity and health care almost non-existent. In addition squatter

(Below left) poor housing in Melbourne, Australia. (Below right) a wealthy house in Delhi, India. (Bottom right) a wealthy house in Perth, Australia. (Bottom left) disease flourishes in places like these shanty buildings in Bombay, India.

settlements are often found on sites nobody else wants because they are subject to flooding, landslips or industrial pollution. Local authorities are often unable to do much about the enormous housing needs, and accept the squatter settlements as a solution to the problem. Services may be provided on a limited scale, but generally there is no sanitation, rubbish collection, water or electricity supplies. There are many pockets of substandard housing in the United Kingdom and Australia, but they are vastly greater in size and number and far worse in quality in most Third World countries.

1 Describe all the services provided to a modern house built in a town or city in the United Kingdom or Australia that are not likely to be provided for houses in a squatter settlement. In what ways will the lack of such services make life more difficult and unpleasant?
2 **a**) Compare examples of high and low quality housing in your neighbourhood or town.
b) Why is it a mistake to suggest quality of housing means the same as quality of living?

Nutrition and health care

In the mid-1980s the world became aware that millions of people in Ethiopia, the Sudan and neighbouring parts of Africa were starving to death. The immediate cause was famine caused by drought and the inability of the African peoples themselves and helpers from all over the world to do much about the problem. Few people realised this was nothing new and not restricted to Africa. The Food and Agricultural Organisation (FAO) and the World Bank suggest that about 500 million people, or more than one in ten on the planet, regularly eat less than the 'minimum critical diet' to stay healthy (at the same time many in the more prosperous parts of the world die or suffer ill-health through over-eating or having the wrong food!). It is calculated that each year some 40 million people die from starvation or diseases related to starvation – that is roughly equivalent to 300 Jumbo jet crashes without any survivors every day! What is striking about this hunger and death from starvation is that it is

very unevenly distributed around the world: more than two-thirds occurs in Asia, and the rest in Africa and Latin America.

People also experience many diseases that are not due to hunger or poor diet, and diseases such as yellow fever, plague and malaria cause an enormous amount of illness and millions of deaths a year. Air-borne, food-borne and water-borne infectious diseases such as cholera and typhoid can result in epidemics on a huge scale. These infectious and parasitic diseases have been eliminated or largely controlled in some countries by the provision of clean water, improved sanitation, health and medical care, and the availability of medicines and drugs, nurses and doctors, clinics and hospitals. However, in many parts of the world such health and medical care just isn't available, and men, women and children suffer needless ill-health, pain and premature death.

As with hunger and starvation, this lack of health and medical care is unevenly distributed around the world. Indeed in many cases the two misfortunes – hunger and ill-health – go together, often with poor quality housing as yet a further disadvantage. There are enormous differences in living standards around the world.

For world map showing levels of nutrition, see page 115

Exercises on page 194
Basic needs in the Third World

3 What relationship can you see between GNP per capita and **a**) life expectancy at birth, **b**) daily calorie intake as a percentage of stated requirements, **c**) infant mortality for the countries in the table?
4 Why don't starving people simply grow more food for themselves?

A comparison of hunger and health in six countries

	Population (millions) 1985	GNP per capita (US$) 1985	Daily per capita calorie intake (as per cent of requirements) 1983	Life expectancy at birth (years) 1985	Infant mortality rate (per thousand) 1985	population per doctor 1980
Low-income countries						
Malawi	7.0	170	95	45	156	41 460
India	765.1	270	96	56	89	3 690
Middle-income countries						
Thailand	51.7	800	105	64	43	7 100
Peru	18.6	1 010	85	59	94	1 390
Industrialised countries						
United Kingdom	56.5	8 460	128	75	9	650
Australia	15.8	10 830	115	78	9	560

1.6 Quality of life

It is hard for anyone suffering from hunger, ill-health or poor housing not to be miserable! However, many people who do have a reasonable material standard of living also lead lives that are unsatisfactory or full of struggle, anxiety and fear. There can be many reasons for this, but before considering some of them it is important to realise that what might be miserable for one person or family could be very satisfying to another – it very much depends on past experiences and future hopes.

Migrant workers going home from South Africa

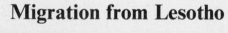

Migrant labour and family life

If it is impossible to find employment near home the only solution (apart from being supported by family or state) is to seek work elsewhere. People have migrated in search of work since the beginning of recorded time, and the process continues today. Migration can be both within a country and between countries.

Sometimes an individual or a whole family will move permanently to a new place. An alternative is where a man or woman leaves home for a temporary period of work in a distant place, sending money home to his or her dependents. In some cases temporary migration is by choice because it provides a salary higher than at home, or for promotion in a job. Far more often, however, there is no choice, and the migrant worker reluctantly leaves home to work away from the people and the place he or she prefers. Through no fault of their own, millions of men and women throughout the world have to become migrant workers and lead lives that may be materially acceptable but unsatisfactory in many other ways. The quality of life of families left at home is also diminished as the price for some material security.

Migration from Lesotho

An example of temporary migration for work is that between Lesotho and the surrounding Republic of South Africa. What usually happens is that between the ages of about 18 and 45 or 50 the male migrant worker moves between his rural home and South Africa on many occasions. He will be away from home and family for anything up to a maximum of two years, and then has to return to Lesotho before signing another contract. The majority of the male migrants work in the mines, and during their stay in South Africa live in mining compounds. Many women also migrate to work in South Africa, usually in domestic service or in factories, and they too have to be away from home and family for long periods. In the late 1970s about 60 per cent of males of employable age worked in South Africa. This was mainly because there was not enough adequate work available in their own country.

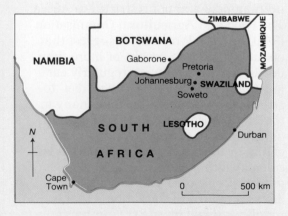

Lesotho and South Africa

Oppression and danger

In many parts of the world the lives of some people are made miserable because of the dislike or intolerance of them by others. This may take the form of being ignored or not allowed to take part in social activities, but can also lead to job discrimination or many social injustices. The most extreme cases of oppression can involve verbal and physical abuse, violence against people and their property, imprisonment or death.

Sometimes this intolerance and hatred is based on differences of culture or race, and when one group has more power than the other, the less powerful can suffer great humiliation and misery if not physical harm. Similar intolerance and hatred can be seen between different religious groups, and even between different sects within one religion. Many religious minorities live in anxiety and fear because of the anger and fury they arouse in the religious majority within their country.

In many countries individuals and groups who hold different political views to those who hold power are persecuted. In democracies attempts are made to tolerate a wide range of views and behaviour, but under other types of government differences of opinion are usually ruthlessly squashed. People who resist are either imprisoned or become refugees by being exiled or escaping the country. A large number, however, accept a great deal of dissatisfaction and unhappiness to avoid the harsh penalties. Throughout the world poor and rich alike are frequently the victims of oppression and lead unsatisfactory and dangerous lives because of racial, religious or political discrimination and intolerance.

1 Name two countries to which UK citizens have permanently migrated in large numbers to seek work, and two countries from which many people have migrated to the UK for the same reason.
2 Describe any examples you know of temporary migration for work of UK citizens to a) other parts of the UK, and b) other parts of the world.

Women in Evin Prison in Tehran, Iran

3 Imagine you are a temporary migrant male or female worker from Lesotho in South Africa. Write a brief letter home to your family, telling them about your job, living quarters and when you will be able to see them again.
4 Give examples from recent news items from a) your own country and b) other countries of the persecution of people because of their culture or race, religious beliefs or political views.
5 Why are governments or individuals with power often very intolerant of people who a) are of a different culture, b) hold different religious views, c) hold different political views?

Exercises on page 195
Disadvantaged groups in India

Political refugees from Kampuchea arriving in Thailand

19

1.7 The Third World

The photographs on this page show two views of the same Third World country, Egypt; the countryside and the city. (Left) the River Nile near El Aiyat and (right) Cairo

The Third World is a name used to refer to a large group of countries to distinguish them from the rest of the world. Like all labels it has to be used carefully, not least because its meaning has changed with time.

During the 1960s it became customary to divide the world into three major groups and to call them the First, Second and Third Worlds. The countries of Europe, North America, Australia, New Zealand and Japan belonged to the so-called First World. They had capitalist economies, and were industrialised, wealthy and powerful. The communist countries of the Soviet Union and Eastern Europe, with their centralised and state-run economies were known as the Second World. The remaining countries of the world, mostly in Latin America, Africa and Asia, were collectively known as the Third World. Although varying a great deal in their environments, size and economies, they were generally poor and powerless, although some contained great mineral, vegetable and energy resources awaiting development. Most Third World countries had been colonies in the empires of the powerful European nations during the first half of the century. Although now independent they were forced to rely on the patronage and aid of former colonial powers or more recently formed powerful nations.

With the huge increase in demand for oil after the Second World War, and the shift in production to countries such as Saudi Arabia and Kuwait in the Middle East, Libya and Nigeria in Africa, and Venezuela in Latin America, the economic power of the world shifted a little. During the 1970s these oil producing and exporting countries became very

The three 'worlds'

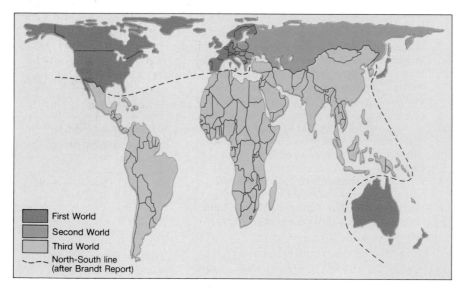

First World
Second World
Third World
- - - North-South line (after Brandt Report)

wealthy and economically powerful, and it made little sense to group them with other countries in the Third World group who remained desperately poor. The pattern of oil production and consumption has changed again, but these countries, most of which have gathered together to form the Organisation of Petroleum Exporting Countries (OPEC), can still be distinguished from others by having this important mineral resource.

Changes have also taken place in the other Third World countries. There has been some dramatic industrialisation in the western Pacific countries of Taiwan, South Korea, the Philippines and Singapore, as well as in Latin American countries such as Mexico and Brazil. Although there remains a great deal of poverty within these countries, as well as problems of enormous international debt, they are in a different category to the very poorest nations that make up the remainder of the Third World. In 1984 the United Nations identified thirty-six least developed countries (LLDCs), the 'poorest of the poor'. There are twenty-six LLDCs in Africa, eight in Asia, one in the Pacific, and one in the Caribbean (see map on page 22). In population they range from the Maldives with 150 000 to Bangladesh with about 90 million. Some of the conditions that lead to countries being included in this unenviable list are given in the extract.

North–South and the Brandt Report

In 1980 an influential group of people under the chairmanship of Willi Brandt, ex-Chancellor of West Germany, produced a report based on a survey of the world's economic and social problems and likely future. The report, and its successor in 1983, was named after the chairman. The Brandt Report used an even simpler classification of the world, as can be seen on the map. It refers to two broad groups of countries, the North and the South, corresponding very roughly to the wealthier and more powerful First and Second Worlds, and to the poorer and less powerful Third World.

All classifications have their weaknesses. They hide enormous differences within the groups and lead to false and

UN to help poorest of the poor

Judged by a range of economic and social indicators, countries in the least developed category show a level of development that is low even by the standards of the Third World.

A high proportion of their population is dependent on agriculture, predominantly subsistence agriculture. Manufacturing output is extremely small and such trade as they carry on consists of one or two primary products. Their climates are often harsh. Literacy is very low and skilled administrators scarce. Health services are minimal and malnutrition often widespread.

On top of this, these countries have been badly affected by the oil price increases and inflation of the 1970s. Falling prices for their exports of primary products, and soaring prices for imported fuels and manufactured goods, has meant that the least developed countries have had to run ever faster in order to stay where they are.

The situation has been made worse by droughts, wars and increasing internal demand for food due to population growth.

Foreign aid has become crucial to most of the least developed countries, paying for more than half their imports in some cases. Although aid has been stepped up, the extra money has been swallowed up by the rapid inflation of prices of vital imports.

The result has been very slow economic growth, or even declining national incomes. At the same time, the population in these lands is rising at about eight million a year.

crude generalisations being made about countries. They also tend to encourage thinking about differences and divisions rather than similarities and interdependence. Nevertheless, when used with care the terms 'Third World' or 'the South' do provide a convenient shorthand way of describing many of the more disadvantaged parts of a very unequal world.

1 From the extract make a list of the main features of the poorest or least developed countries (LLDCs) of the Third World.
2 Draw a statistical chart or graph to show the following statement: 'The North has a quarter of the world's population and about four-fifths of its income. The average person can expect to live to about seventy years, rarely go hungry, and have an education to at least secondary level. The South has three-quarters of the world's population but only one-fifth of its income. The majority of people have a life expectancy of fifty years, many suffer from hunger and malnutrition and about half have little chance of becoming literate.'
3 Use the map of GNPs per capita (page 22) and the graph you have just drawn to explain the dangers of generalising about the North and the South.

Exercises on page 196
The least developed countries

1.8 Summary

The first chapter of this book has introduced some ideas about the differences and inequalities to be found in the world. Before going on to consider some of the explanations that are given to account for these differences, and to see how these apply in some Third World countries, these ideas can be brought together in a summary.

☆ There are great differences in ways of life, living standards and quality of life around the world. Some of these differences contribute to the richness and pleasure of everyone's experience, but others reflect injustice and misery, and need to be changed.

☆ Figures for economic performance and wealth of countries can give a simple measure of difference, but they ignore other important indicators of well-being such as health care and diet. They also mask great differences within countries.

☆ Our knowledge and understanding of the world is conveyed through a variety of media – words, pictures, maps, diagrams, statistics, and so on. Even with the best of intentions these are bound to be biased and give only a partial picture of the real situation. In some cases bias is deliberate in order to stress a special viewpoint. Therefore, all data and images need to be treated with caution.

☆ No individual, family or country exists in isolation. All are influenced by the wider world. Individual people and countries have some control over their lives, but are also dependent on the behaviour of others. The economic and social development of a country may depend on the actions of other nations, as well as on self-help.

☆ Development is ultimately about individual people, the quality of their lives and their relationships to others.

Gross National Product per capita and the location of the world's least developed countries (LLDCs)

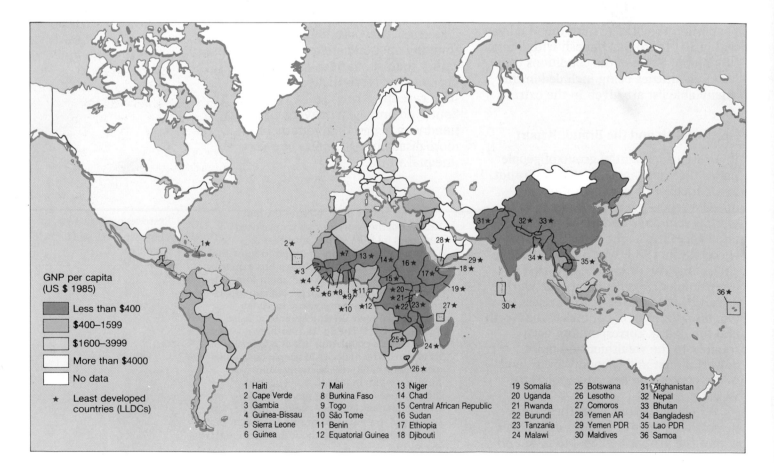

GNP per capita
(US $ 1985)

Less than $400

$400–1599

$1600–3999

More than $4000

No data

★ Least developed
    countries (LLDCs)

1 Haiti	7 Mali	13 Niger	19 Somalia	25 Botswana	31 Afghanistan
2 Cape Verde	8 Burkina Faso	14 Chad	20 Uganda	26 Lesotho	32 Nepal
3 Gambia	9 Togo	15 Central African Republic	21 Rwanda	27 Comoros	33 Bhutan
4 Guinea-Bissau	10 São Tome	16 Sudan	22 Burundi	28 Yemen AR	34 Bangladesh
5 Sierra Leone	11 Benin	17 Ethiopia	23 Tanzania	29 Yemen PDR	35 Lao PDR
6 Guinea	12 Equatorial Guinea	18 Djibouti	24 Malawi	30 Maldives	36 Samoa

Chapter 2

Environments, resources, and development

Are harsh and difficult environments a cause of underdevelopment? How does poverty force people to destroy the very environments and resources on which they depend for survival? When the resources of such environments are developed, does this lead to an improvement in living standards and quality of life for the local peoples? How can resources and environments be developed for present essential needs without destroying them for future generations?

Living in a harsh environment:
a farmer winnowing grain in
the Tigre region of Ethiopia

2.1 The drylands and food production

There are five large areas of desert in the world that are so arid they cannot sustain plant growth for food or animal grazing. Some people live at a few oases, where underground water is available, near irrigated strips alongside permanent rivers and in settlements established to obtain oil or other mineral wealth, but otherwise there are no permanent inhabitants of the true deserts. These harsh environments, perhaps underlain with rich resources but almost empty of people, occupy about 6 per cent of the earth's land surface.

On their margins are larger areas of semi-desert. There is some rainfall, and although it is low in total and unreliable in occurrence it does provide some limited grazing land and a chance to grow a few crops. These semi-arid lands occupy a further 28 per cent of the earth's land surface, and are the homelands of hundreds of millions of people. A good example is the Sahel, a zone extending across Africa to the south of the great Sahara desert.

Sahel pastoralists

The Sahel is a wide and varying zone. Its rough position is shown on the map. In the drier area near the desert margin the people are mainly pastoralists living from their cattle, sheep and goats. In the rainier parts to the south there are settled farmers living in villages and some people working and living in the small number of towns.

Because rainfall is so low and unreliable, many pastoralists lead a nomadic way of life, moving to search out areas of grazing for their animals. Traditionally they have spread over large areas because they need lots of space to avoid overgrazing. During the drier season, between about March and September, these pastoralists move south to the farmed areas. They are able to exchange animals and milk for grain and other needs. Their animals graze on the fields from which crops have been harvested, eating the stubble and adding manure to the soil. Water is available from the wells in this southern zone.

In a year when the rains arrive, the pastoralists move off to their traditional grazing lands and build up their stock again. If there is a delay in the rainfall or it is only slight, they may have to sell off some of their stock and stay nearer the settled areas.

Sahel cultivators

Most of the Sahelian cultivators are subsistence farmers, hoping to grow enough during the year to feed a family and dependents. If the harvest fails they have little money to buy anything from the market. Their main food crops are millet and sorghum, maize, beans, and

Pastoralists in the Sahel. (Below) the Fulani, like this young herdsman, are nomadic pastoralists who move with their cattle within the Sahel and, increasingly, in lands to the south. (Bottom) a Tuareg encampment

rice if near rivers that can provide irrigation or flood water for their fields. Provided the rains come, and there is some fertilizer for the soil, and the fields are not cultivated too often, a reasonable living is possible. But if the rains fail, or the land is cultivated without time for the soils to recover, or fertilizer is not added, then the situation becomes very precarious.

Some farmers are more prosperous and, because of their greater wealth, are able to own more land and use more effective farming methods. They may be able to use irrigation more easily, buy fertilizers and pesticides and get grants from agencies and governments for experimental work. One reason for their wealth is that they grow cash crops that are exported to other parts of their country or abroad, such as groundnuts for use in the manufacture of vegetable oils and cotton for the manufacture of textile goods. They are encouraged to grow such export crops by grants, loans and guaranteed prices offered by governments, who then gain income through taxes. The problem is that it takes much land that could be used for food production and uses up aid that might be put to improving the farming practices of the subsistence farmers. Women farmers, who have land and responsibilities in the traditional system, become just labourers on cash crop farms.

Life in the Sahel has always been precarious, depending as it does so much on unreliable rainfall. When numbers

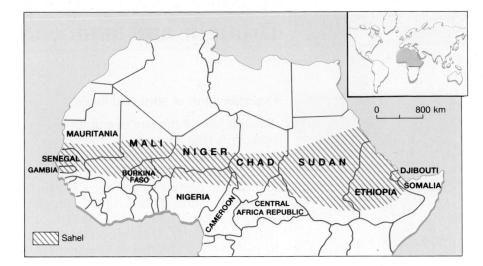

The countries of the Sahel

were small, and pressure on both crop and grazing land not too severe, the Sahelian people were able to have a reasonable living standard and live in ways they have enjoyed for centuries. However, some of the changes that have occurred during recent decades, while improving output and wealth for some, have put new strains on the delicate balance between people and environment.

Exercises on page 197
The climatic conditions of northern Africa

1 a) Name and locate the true desert areas of the world. b) Explain how life is possible in a desert oasis, riverine desert village and desert oil or mineral settlement.
2 Describe or sketch and label the significant features shown by the photographs of the Fulani pastoralist, the Tuareg encampment, and the riverside village in Niger.
3 What do you think would be the greatest anxieties and pleasures experienced by a Sahelian pastoralist or subsistence farmer?

(Below right) a village on the banks of the River Niger, Niger. (Below left) a fertile market garden on the banks of the River Niger, Mali

2.2 Drought and famine

Desertification of semi-arid lands

A major fear of the late twentieth century is that some environments are being damaged to the point of destruction. The semi-arid lands are being turned into deserts by a process known as desertification. This is threatening the living standards and livelihoods of millions of people. Desertification might be the result of changes in climate over thousands of years, but it is happening so quickly and widely that human actions are probably the main reasons. The harmful effects of man's activities are, however, most severe during periods of prolonged drought.

One cause of desertification is overgrazing of the natural vegetation by cattle, goats and sheep. Pastoral farmers must have grazing land in good condition for their animals to survive. The combination of loss of potential grazing land to the cultivators and the poor quality of vegetation due to lack of rainfall means that too many animals are trying to survive on too little pasture. The overgrazed vegetation dies and

begins the process of desertification. In addition to this enormous quantities of woodland are destroyed for fuel because no other source is available. Yet more woodland is lost as subsistence farmers clear it for cultivation.

A further reason is overcultivation of the land. In some cases this is done quite ruthlessly to make quick profits from the sale of cash crops or food in local or overseas markets. In other cases subsistence farmers try and get more from the soil than is possible, largely in order to feed themselves and their families. But without leaving the fields fallow for a while, or allowing animals to graze and manure them, or buying and adding fertilizer, the soil becomes unproductive or lost through erosion. Even if irrigation can be afforded there is a danger of sterilisation of the soil by salts unless it is carefully managed.

Sometimes through desperate need, but often through ignorance or indifference, large areas of semi-desert are being transformed into desert incapable of supporting pastoral or subsistence farmers.

Drought and famine in the Sahel

In each of the years between 1969 and 1973 the rains failed to arrive and the resulting drought had a tremendous impact on the people of the Sahel. It is impossible to tell just how many thousands of people died from starvation or related diseases. Millions of animals died. Vast amounts of aid were given to the region, and when the rains returned it was expected that the lessons had been learned and that such a tragedy would never occur again.

However, the rains failed repeatedly during the late seventies and early eighties. The impact was even more horrific. People from all over the world who had hardly heard of Mali or Sudan or Ethiopia were shocked by the television pictures of mass misery, and many gave to solve the immediate problem of starvation and death. Some idea of the

A poor millet harvest

For world map showing drought and famine, see page 113

Tuareg nomads in western Mali rely on desert wells like this

situation is given in this extract describing a scene repeated in hundreds of places across the Sahel:

'By December 1984, about 25 000 people were gathered in feeding camps around the ancient trading centre of Gao in Mali, 200 miles downstream from Timbuctoo on the River Niger. All were destitute. It was bitterly cold at night and few people had blankets. The cholera epidemic along the banks of the river had died down by then but measles was rampant. The cemetery was full and each day more bodies were being buried in the sand just outside the city. All over the country poor peasant farmers and hungry nomads were leaving their homes in desperation...'

The most immediate need was to provide food, water, shelter and medical care to the hungry and dying, and in the face of many difficulties great efforts were made to do so. However, the longer term aim remains to rehabilitate the Sahel. No-one can control the rainfall. Many other ideas have been suggested and put into practice, but they each bring their own problems and the size of the task is enormous. The Sahel, like all semi-arid environments, can enable a limited number of people to live in a way they enjoy and without fear of starvation, but it is a precarious balance between environment and people, and once that is upset the standard of living and quality of life deteriorates dramatically. Developments to improve the quality and security of life of the great majority of Sahelian people have to be carefully thought out and managed.

1 Why is it impossible to tell how many people died across the Sahel during the famines of the early seventies and eighties?
2 Write a short article under the heading 'Desertification – a man-made disaster?'
3 Read the article (right). **a)** State anything you find surprising about the life-style, customs and attitude of the Fulani woman, Guntu Mamane.
   b) In what way are 'seeds of a new life' being sown in this project?

Exercises on page 198
Environmental damage

Members of a grain co-operative recording sales

Sowing seeds of new life in arid sands of Niger

The hands of a Fulani woman should be like silk, lamented the stately nomad. Guntu Mamane held out her hand to display the roughened skin and callouses.

For the past 50 years she has wandered the Sahara with her small family tribe living off the milk of her camels and goats. But last year, in the greatest drought in northern Niger this century, she lost eight camels, two donkeys, eight goats and eight sheep.

Since then she has for the first time in her life been forced to use a hoe. 'I had nothing to live on and I had two children to look after. I am married to an old man who has no teeth and cannot see.' The teeth are important. They are the mark of beauty in a Fulani man and it is important for a woman to have a beautiful husband for it is the women who choose their mates at the annual dancing ceremony.

'I have to give him food and clothes too,' she said, with cold charity.

Last year more than 700 000 Tuareg and Fulani nomads lost all their camels and cattle in Niger.

As milk and cheese are almost their entire diet, the death of the livestock meant an end to their traditional way of life.

About 300 000 of them went to camp on the edges of towns in and around the Sahara but about 400 000 were taken into a new government programme which uses purpose-drilled boreholes and hand-drawn irrigation to grow crops in the dry season in the very sands of the desert.

Many times a day Guntu draws water and pours it around the stems of the dozen rows of millet and sorghum she has planted. The crop is not a good one. In the shade the temperature is 115 degrees F. In the sun the heat is penetrating and the ground soaks up the water so rapidly that within minutes it looks as though it has never been moistened at all.

But she has saved enough so far, because of the high prices of grain, to buy two new goats. She will carry on planting until she has enough to form a viable herd; and then she will move on.

2.3 Arid North Africa

North of the Sahel lies the enormous Sahara desert, extending from the Atlantic Ocean in the west to the Red Sea in the east. It consists of vast stretches of rock and sand desert, with occasional mountain ranges, and the only permanent river to cross it is the Nile which flows northwards through Sudan and Egypt. With the exception of a few oases and their settlements the Sahara presents a harsh and inhospitable environment. It makes up the southern parts of all the African countries fringing the Mediterranean Sea.

Oil from the Libyan desert

The country that is now Libya was once an Italian colony, taken by force and settled by over 90 000 people. These colonists built towns, harbours and roads and began to develop the resources of the country, but the Arab peoples were savagely treated. Following the defeat of Italy in the Second World War, the United Nations took control of

Libya

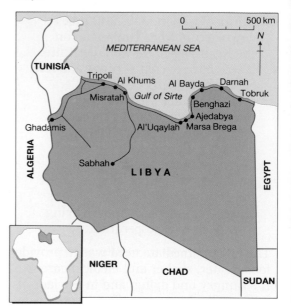

An oasis in Algeria

the territory, and in 1951 it was declared an independent kingdom. At that time the World Bank described it as the poorest country on earth. Apart from the towns of Tripoli and Benghazi and smaller settlements along the coast, and oases where dates, olives, almonds, figs and grapes were grown, most people lived a desperately poor life of subsistence farming. A big change came with the discovery of oil. Foreign companies were first allowed to explore in 1955, and four years later the first discoveries were made in the desert south of the Gulf of Sirte. The first oil was exported in 1961, and Libya soon became a major exporter and earner of money from the industrialised nations of the world.

In 1969 a group of young army officers forcibly took control of the country. Under the leadership of Colonel Gaddafi the Revolutionary Committee determined to improve the living standards of the Arab people and create an Islamic socialist state. Because there remains a shortage of Libyans experienced in oil technology the country still uses the expertise of foreign oil companies, but production is now under the control of the state and the country as a

Burning waste gases in the Libyan oilfields

whole benefits from the sale of oil. In the early 1980s oil and natural gas provided almost 100 per cent of exports and 60 per cent of national wealth.

Using the wealth from oil

Libya's population of only about 3.1 million people is small considering its great size. However, much of the wealth from oil has been used in attempts to make the country self-sufficient in food production – in an environment where less than 1 per cent of the total land area was considered suitable for crop growth! This has been done by improving or reclaiming badly farmed or abandoned land, by importing farm machinery and fertilizers, by rebuilding old water projects and harnessing newly-discovered underground water supplies, and by resettling people on modern farms and in rural villages provided with new roads and services. At the same time new industries have been developed, based on oil and natural gas resources, harbours, roads, airports and railways constructed, and many new factories built. All this has been possible because of the oil wealth.

Alongside these economic developments there have been striking improvements in housing, education and health care. Special attention has been given to the rights of women in all these developments. Even with all this expenditure there is money to spare, and much of this has been spent on buying military equipment, giving aid to friendly foreign nations and groups and investing in foreign companies.

Many problems remain, including the sheer time it takes to introduce so many changes and for them to have an effect. One is the continuing lack of local labour and skills in spite of the expansion of education and training. In the early 1980s it was officially estimated that foreigners from all over the world made up more than 18 per cent of the population and 45 per cent of the workforce. Many Libyans are accused of not making enough effort to produce the Islamic socialist state planned by their leaders, while the country has a bad reputation for the harsh and brutal treatment of political opponents. Nevertheless there seems little doubt that the majority of Libyans have a far higher living standard now than before the discovery of oil and the rise of the Revolutionary Committee.

1 Read the extract and explain a) where money and help for the project was coming from b) who objected to the scheme, and why, and c) why foreign banks are reluctant to make loans to Libya. Why did the USA proclaim and try and establish an international economic blockade in 1985?
2 Describe some of the a) economic and b) social developments in Libya. Why do some people argue that the living standards in Libya cannot be properly described as development of the society?

Modern irrigation methods in Libya; one of the benefits of oil wealth

Exercises on page 199
Niger

Gaddafi launches scheme to tap desert

Colonel Gaddafi has inaugurated a big project to pipe underground water from the desert to cities and farms thousands of miles away.

There were cheers from Bedouins and American businessmen as Gaddafi launched the plan, one of the most ambitious and expensive in Libyan history. The finished project will extend 2400 miles.

The final cost could be $7bn but there is doubt whether Libya can raise the cash because of dwindling oil revenues and a reluctance on the part of foreign banks to provide loans.

Egypt and Sudan, Libya's enemies and neighbours, are concerned that the underground water feeds the Nile and the project would deplete its waters.

For the launch Libya flew in Western journalists, diplomats and foreign businessmen connected with the five-year project. Dignitaries included officials of the Texas company of Brown and Root, the overall consultant, and the Ohio company of Price Brothers, which will oversee construction of the pipe.

Asked why American firms were aiding Libya, considering its bad relations with the United States, Mr Gayle Price Jr of Price Brothers, replied: 'This is not a military project. It is humanitarian.'

29

2.4 The Arctic

Although there have been dramatic changes in recent decades, the remote and ice-bound polar environments remain amongst the least disturbed on earth. The Arctic is an enclosed sea, the smallest of the oceans, and almost entirely landlocked. Much of it is covered with ice throughout the year, although the margins of the ice move southwards during the winter and move back towards the North Pole as the ice melts during the summer. About one-third of the Arctic Ocean is underlain with the continental shelves of the surrounding continents, and here are found many of the world's richest fishing grounds. The most northerly parts are regions of perpetual snow and ice.

This polar zone merges southwards into the tundra which in turn extends as far as the coniferous forests encircling the Arctic lands in North America, Europe and the USSR. During the winter the tundra is bleak and bitterly cold, and covered with snow and ice. In summer, for a brief period only, daytime temperatures rise above freezing point and the surface soil thaws; the sub-soil remains frozen, creating large areas of water-logged marsh and bog. Elsewhere the summer provides many flowering plants, mosses and lichens to complement the scattering of dwarf shrubs and stunted birch and willow trees. The sun can shine for long hours in the summer unless obscured by cloud, but it stays low in the sky and therefore provides relatively little warmth. At the Arctic Circle on midsummer's day the sun remains above the horizon for 24 hours. The number of days with a 'midnight sun' increases towards the pole, where the sun doesn't set for six months. On the other hand, the winters bring days when the sun doesn't rise, adding the difficulty of darkness to the severe cold. There is, however, a surprising amount of wildlife in these tundra lands, and this has enabled people to live in the harsh environment for many thousands of years.

Traditional life in the Arctic

Few people live in the true polar zone of perpetual ice and snow, but in the marginal tundra lands people have used great skill and ingenuity to survive on the resources provided by fish, seals, whales, polar bears and caribou and on the limited berries and summer vegetation. From the animals they obtained food, and from their skins, sinews and bones made tools and weapons, clothes and footwear, kayaks and sledges. With the help of dogs equally adapted to the environment they were able to move about the snow-covered landscapes. A whole way of life was developed by the North American Inuit or Eskimo people and the Arctic Indians. To most outsiders it was hard and difficult, but the

A modern Inuit hunter works with a power sled and a rifle

indigenous or local peoples had learned to live in harmony with their environment.

Similar ways of life existed in the Arctic lands of Asia and Europe. The Lapplanders of northern Scandinavia (who refer to themselves as the 'Sami') were traditionally dependent on semi-domesticated reindeer, and perhaps fish if they lived in coastal areas. In winter they lived nearer the tree-line in wooden huts and tents in sheltered valleys, following the herd as it searched for moss under the snow. In summer the herds moved north to summer pastures nearer the coast.

These various peoples of the Arctic lands traded with outsiders, exchanging skins and meat and traditional clothing for paraffin, salt, tobacco, alcoholic drinks and weapons. As communications improved the governments of the territories became more interested in the remote indigenous people, sometimes in a genuine effort to help them, at others to try and bring them under some form of control. As settlements were built and contact increased so the local peoples were influenced by the goods, behaviour and beliefs of the outsiders. There is no doubt that the material standards of living of many Inuit, Indian and Sami, for example, improved a great deal as a result of this contact and the ability to buy and use modern equipment, food and luxury goods.

Many also received formal education and training for the first time. In this sense the communities became more developed. But there was a price to pay as old customs and traditions changed, and freedoms were curtailed. In 1977 more than 400 leading Alaskan and Canadian Inuits and Indians, Native Greenlanders and Scandinavian Sami met to launch the Inuit Circumpolar Conference for all Arctic peoples outside the USSR. Amongst other aims they want to arrange legal and economic frameworks to ensure their children and grand-children might continue the traditions and beliefs of their ancestors. As we shall see on the following pages recent developments have had an even greater impact on the indigenous peoples of the Arctic lands.

The Sami (or Lapp) people of Scandinavia are dependent on reindeer herds for their livelihood. Their way of life was severely affected by the build up of radioactivity in their reindeer following the nuclear disaster at Chernobyl in the USSR in 1986. The reindeer grazed contaminated moss and lichen

Exercises on page 200
The Canadian Arctic

Arctic lands

Chukchi
Inuvialuit
Yakut
Nenet
Sami
Inuit
Kutchin (Arctic Indians)
Permanent ice
Maximum extent of sea ice

1 Name all the countries that have territory in the Arctic lands, and for each name an indigenous group of people who have lived in the area.
2 Describe all the benefits that a) traders, and b) governments might bring to these scattered communities. What are some possible disadvantages?
3 In what sense were the peoples 'underdeveloped'?

2.5 Development in the Arctic

Economic development

Following the discoveries of vast oil, natural gas and mineral deposits in the Arctic lands, major companies have invested huge sums of money in developing these resources. Towns have been built in the once remote places where the oil, gas and minerals are being exploited. Roads and pipelines and oil tanker terminals have been constructed to take the resources out of the Arctic lands to where they are needed. Large dams and power stations have been built in these bleak regions to provide the necessary energy for the mineral operations and the towns.

The search for, and exploitation of these resources, has led to the leasing, buying or taking of the traditional hunting and grazing grounds of the indigenous people. This has dramatically affected their ways of life and the environment. Transport of oil and gas affects the environment: huge tankers that force a passage through the ice have a damaging effect on the fish and whale populations and provide a threat to the seas and coastlands from spillage of oil and the resulting pollution. Economic development usually brings great benefits to the developers, and unquestionably to some of the local people. But as we shall see there can also be unwanted harmful effects to the environment and the traditional societies of the Arctic.

Military development

Another cause for major change is the fact that the two superpowers, the USA and the USSR, feel the need to defend themselves from attack from each other. Although the two nations are only a short distance apart in Eastern Siberia and Alaska, the main centres of population are most accessible over the polar region of the Arctic. So to prevent attacks by aircraft or missiles a vast network of monitoring stations and military bases have been built by both nations in a circular zone around their northern frontiers or those of their allies. For military and defence reasons alone many remote areas of the Arctic lands have seen the construction of large settlements and bases. Governments can be determined and ruthless about taking lands on the grounds of national security. While the bases and stations may lead to more jobs and improved communications, they too can bring environmental and social disadvantages.

The SS *Manhattan*, a nuclear powered oil tanker specially constructed for use in the Arctic

A Siberian gas pipeline

Impact on the native peoples

Some 2 million people now live within the Arctic Circle, of whom about 800 000 are descendants of the nomadic communities whose hunting and survival skills kept them going in these harsh environments for at least 10 000 years. But life has changed for most of them because of the activities of the economic and military developers and environmentalists who have interfered with their traditional ways of hunting to protect the animal species.

In material terms the living standards of many are better than they used to be, but the communities are worse off in other ways. There are too few jobs in the new industries, and these are often taken by outsiders seeking to earn big incomes quickly. Men and women unable to hunt for a living or trade for goods they need are forced to rely on social security. There is a pattern of boredom, excessive drinking and lawlessness plaguing many of the communities, and many wonder what life can offer their children now that traditional ways of life have gone.

Some groups are helped by their governments. Compensation for loss of land has gone into building schools, hospitals and other public buildings or providing electricity, safe water or satellite telephones. Some Inuit have become successful businessmen in banks and the hotel industry. In Canada, unlike in Alaska, the income from sale of traditional land is held in trust for the future collective benefit of the communities as a whole. In the USSR the main ethnic and cultural groups are listed separately in the census, but otherwise treated as ordinary citizens with no special rights and no claim to land, either as individuals or communities. Many try to retain their traditions while being Soviet citizens.

Many resources are being developed in the Arctic, and people from outside the region frequently earn high incomes from working there. Governments and companies also benefit from these developments. But it is not easy to argue that the lives of the traditional Arctic peoples have changed for the better.

1 **a)** Why have the mineral resources of the Arctic region been developed? **b)** Do the Arctic peoples have any right to benefit from the developments if other people's money has been used?

2 With reference to this and the previous section discuss the statement 'The Arctic peoples are less developed now than they were before the arrival of the developers'. Is it possible for any outsider to say whether the native peoples are better or worse off?

The map shows how military bases and early warning systems circle the Arctic. The photograph shows the US air base at Thule, north-west Greenland

Exercises on page 201
Inuit life-styles and their hopes for the future

33

2.6 Wealth from Antarctica

In many ways the polar wastelands of Antarctica are like those of the Arctic – dramatic icy landscapes, inhospitable and dangerous. But whereas the Arctic is an ice-covered enclosed sea, Antarctica is an ice-covered land mass of continental size. The vast mountainous island, centred over the geographic South Pole, is twice the size of Western Europe. Its ice-cap, averaging 2300 metres thickness, contains nine-tenths of all the ice on the planet. If it melted and added its water to the oceans all land that is less than 65 metres (200 feet) above the present sea-level would vanish beneath the new seas!

Unlike the Arctic, it has only two flowering plant species and a number of mosses and plants. It has no vertebrate animals, and no long-established settlements of native peoples. The surrounding oceans, however, provide one of the most productive zones on earth. During the summer masses of microscopic materials well up from the ocean floor enabling phytoplankton to flourish. These provide food for the millions of tiny shrimp-like krill, which in turn provide food for fish, whales and about forty species of bird. These birds include the albatross and the huge numbers of penguins.

The photograph shows the Antarctic landscape; 'Nunataks' by the midnight sun. National claims on Antarctica (below) and a cross-section through Antarctica (bottom)

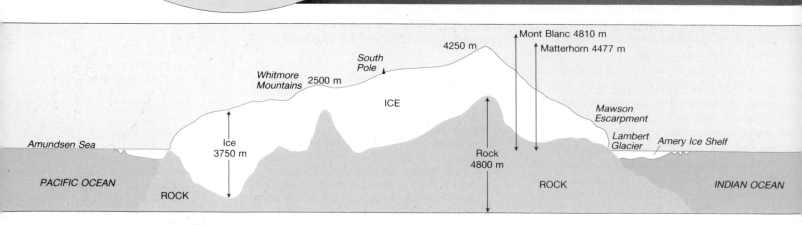

Food and mineral resources

Massive swarms of krill are trawled from the Southern Ocean by Japanese and Russian ships, and the catch processed to make animal food. While there is a wealth of plankton, there is not the same wide continental shelf as in the Arctic to support enormous shoals of fish. Only a few species of fish such as the Antarctic cod and the Patagonian hake are widely caught, mostly by Russian, Polish and East German fleets. Whales have long been hunted in the area. In 1986 the International Whaling Commission introduced a world-wide ban on commercial whaling, but some countries continue, and have also taken up 'scientific' whaling for research purposes, on which there are no restrictions. Some species are still in danger of becoming extinct.

The oceans could one day provide the people of the world with a rich and plentiful source of food, but more interest is being shown in the probable mineral wealth of the continent. A wide range of minerals have been detected, and although exploiting anything from this remote and inhospitable environment would be difficult and costly, they would prove of enormous value in times of mineral and energy shortage.

Scientific exploration

Although there have never been permanent settlers in Antarctica similar to the Inuit of the Arctic, there were in the mid-1980s about 1000 people living there temporarily. They were mostly scientists and support teams living on specially prepared bases. Thirty years previously these bases were on the coast because the interior was inaccessible, but the research stations were then gradually built far inland, including one at the South Pole itself. These scientists come from many countries and they work in harmony on their research projects. Most are relatively young, and they stay about two years at a time. Often they are engaged in routine and boring jobs at base, but some go in small teams to more remote places, often flown by ski-equipped aircraft.

Who owns Antarctica and its wealth?

Seven countries claim to have historic rights to parts of Antarctica, as shown on the map. Some of the countries are in dispute, and generally the claims are not recognised anyway. Following scientific co-operation during the International Geophysical Year in 1959, the seven nations with territorial claims, plus the USA, USSR, Japan, Belgium and South Africa, signed the Antarctic Treaty. This meant that until 1991 they would not pursue their individual claims to parts of the continent, but share in scientific and exploration work. Signatories to the treaty were increased with the joining of Poland, West Germany, Brazil, India, China and Uruguay. These 18 countries comprise about 80 per cent of the world's population, but the remainder feel they too have a right to share in the continent's probable wealth. Some environmental groups argue that Antarctica should be designated for ever as a 'World Park', free of all commercial, military and environmentally damaging activity. The difficulty is, if mineral resources are proved to exist in large quantities then people will want to use them – but who has the right to do so?

1   a) Which countries would have most right to ownership of parts of Antarctica on the basis of nearness? b) How was it that some countries very far away were able to claim 'historic' rights of ownership?
2   Argue the case a) for, and b) against the general principle of developing the resources of this last remaining unexploited continent.
3   On behalf of the 'have-not' nations without a stake in Antarctica, Malaysia argues that the continent is 'the common heritage of all nations of this planet'. Say whether you agree or not, and why.

> **Exercises on page 202**
> The Antarctic Treaty and the issues surrounding it

Members of a Brazilian scientific research station in Antarctica indulge in their favourite pastime

2.7   Mountain environments

Large areas of the earth's surface consist of high mountain ranges and enclosed, high altitude valleys. Despite the great difficulties and hazards, mountain environments in many parts of the world have been settled for centuries. However, the rugged landscape, high altitude and associated harsh weather combine to make economic development difficult, and this is reflected in the generally low standards of living. These mountain communities experience ways of life very different from those of arid and polar regions.

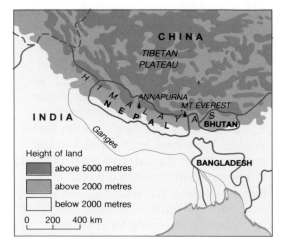

Nepal, Bhutan and surrounding countries

A view into the Manang Valley, Nepal (below), and a house in the village of Manang (bottom)

Life at high altitudes in the Himalayas

One of the most spectacular of these mountain regions is the Himalayan range in Asia. Its huge snow-capped mountains stretch in an east–west curve across central Asia for over 2400 kilometres. To the north lies the vast, bleak plateau of the Chinese province of Tibet. To the south lies India and the land-locked countries of Nepal and Bhutan. The range includes some of the greatest mountains in the world, including the highest, Mount Everest. There are no permanent settlements in the severe environments of the mountain peaks, but thousands of villages are scattered in the more sheltered valley sides and floors immediately below.

Until the middle of this century much of the Kingdom of Nepal was closed to foreign travellers, and it was not until the 1950s that outsiders were able to explore some of the more remote border areas. One such area of restricted access lay west of Kathmandu (the capital city) and north of the massive Annapurna range. High in the mountain valley, only a few kilometres south of the Chinese border, a number of villages still retain a traditional lifestyle.

The village of Manang stands about 3500 metres above sea-level, at the head of the Marsyandi Khola valley. The tightly-clustered flat-roofed houses made of stone and timber are built in tiers against the hillside. They overlook the fields, also distributed at different levels, in which the villagers grow barley, buckwheat and potatoes. Besides this farm produce, they rely on their goats, sheep, oxen and yaks. These provide food and materials for their own use and for trade. The villagers trade with visiting merchants and other settlements, both lower down the valley and in the south of the country.

The villages and land in the area used to belong to the rulers of Tibet. Links with Tibet explain the influence of Buddhism and the existence of temples; the surprisingly large number of lamas (Buddhist priests) perform religious rites and also contribute to farming in the villages. Further south the influence of Hinduism is far stronger.

In the clear air and bright sunlight of summer, life in such villages may seem quiet and remote, but the communities face many hardships. The very remoteness that permits the peace and quiet means that communications are difficult. Roads and motorised traffic are few, and much carrying is done by yak or human porterage. Many natural threats and disasters are common in mountain environments. An early snowfall can ruin a year's entire crop, while a severe winter can kill off large numbers of livestock. Although experience has shown which sites are dangerous there are often avalanches, landslides and earthquakes to threaten the villages, while in some places flooding of rivers is an additional hazard.

It is hardly surprising that material

Wealth and living standards compared

Country	Area (thousands of square kilometres)	Population (millions) 1985	GNP per capita (US$) 1985	Percentage of rural population below absolute poverty level 1977–84	Life expectancy at birth (years) 1985	Urban population (as percentage of total population) 1985
Nepal	141	16.5	160	65	47	7
Bhutan	47	1.2	160	65	44	4
United Kingdom	245	56.5	8 460	–	75	92
Australia	7 687	15.8	10 830	–	78	86

standards of living are not very high, and that there is need for more education and health care, as well as technical and financial help to improve production in the region. High altitude environments can present particular problems for economic development.

Exercises on page 203
Mountain areas

1   a) Look carefully at the photographs of housing and farming activity at Manang. What clues do they provide about the living standards of the people? b) Why is it dangerous to use the statistical data provided in the table to describe living standards in these mountain villages of Nepal?
2   List some of the ways in which high altitude mountain environments lead to relatively low standards of living for local peoples. In what ways might these people enjoy life in the area in spite of the difficulties? What would you like and dislike about living in such an environment?

Cultivating barley on the edge of the village of Manang

2.8   Change in the Himalayas

Many changes have taken place amongst the communities living in the Himalayan foothills and mountain slopes over the past half century. These are well illustrated in the northern zone of Nepal.

The Gurkhas

For over 150 years there has been a tradition for young men from Nepalese villages to join the British Army. These Gurkha boys still compete fiercely for the chance to join the regiment, partly for esteem, partly to escape the poverty of the area and partly to earn money for the family back in the village. Every year about 80 000 try their luck when the first stage of recruiting starts in the villages. Many get no further than this stage, while the successful ones have to go through yet more selections. Only the strongest, fittest, largest and most able are finally accepted. They can then look forward to adventure, travel, high status at home and money for themselves and their families. At the end of their service they will have seen ways of life very different from that in their mountain villages, and this will affect their attitude to change and development.

Gurkhas on parade, Nepal

Improved health care

Other changes result from contact with the Nepalese of the lower valleys and plains to the south and with people from other countries. The building of roads means that many places are much more accessible than they used to be – bringing both advantages and disadvantages. Education and health care has greatly improved, but there are many scattered parts of Nepal where leprosy remains widespread. This is a crippling and disfiguring disease that is hard to detect in its early stages and hard to cure. In 1975 the government set up the Leprosy Services Development Board to protect the healthy population and reduce prejudice against the disease. Teams of paramedical workers seek out cases and provide health education while treatment is obtained from government 'basic health posts' and mission hospitals.

Expeditions and tourism

A different sort of development has resulted from the existence of Himalayan peaks within and on the borders of the country, such as Annapurna and Mount Everest. Following early explorers and missionaries into the area, climbing expeditions began to make the Himalayas widely known to the outside world. Since the 1950s mountaineering and mountain walking expeditions, and general tourism, has become a feature of the region. On the one hand this provides employment as porters and brings income to the guest houses and small hotels that cater for people from all over the world. On the other hand there are some harmful consequences.

Tourists everywhere tend to generate rubbish and waste as well as wealth, and this is certainly the case in the Himalayas. At the Mount Everest Base Camp so much waste has been buried or left around that local water is too polluted to be safe for drinking. South Col, at 8 000 metres, has such a litter of discarded oxygen cylinders, ice-axes, crampons and mountaineering gear that it has been described as 'the world's highest junkyard'. Much more dangerous in the long run has been the felling of trees and woods from the hillsides. Torrential rain falls on these slopes during the July to September monsoons, and the soil on exposed slopes gets washed away and lost. Well over half of Nepal was forested in the middle 1940s, but by 1980 this had been reduced to a quarter. Much of this is due to increased population pressure and local demands, but there is no doubt that the big influx of visitors adds to the environmental damage. While in some senses these changes can be considered 'development' and of value to local people, in others they are a threat to their living standards and quality of life.

1 Draw an annotated sketch of the Marsyandi valley to indicate all the features mentioned in the last four pages.
2 List the advantages and disadvantages of recruitment of young Gurkha men into the Brisith Army from the local villagers' point of view.

The valley of the River Marsyandi

Sherpa porters. The word 'Sherpa' comes from a Tibetan dialect and means 'Easterner', but it is now commonly used to refer to any high altitude native porter in the Himalayas

Exercises on page 204
Tourism in the Himalayas

39

2.9 Hot, wet environments: the rainforests

The climate and weather of places in the band encircling the earth around the Equator are very different from that experienced in Britain and other mid-latitude countries. It is always very hot and rainfall is always heavy. This means the air is always full of water vapour and the humidity is high. Throughout the year there is little variation in the pattern of weather and very little difference in the length of day and night.

Climate has a very big influence on plant cover and the animal and insect life that can live in the habitat. The interrelationship between climate, plants and living creatures produces a particular ecosystem, and that of the rainforest is quite remarkable. A green band of rainforests extends to about 10 degrees north and south of the Equator. A huge variety of plants make up the luxuriant vegetation which in many places has an entirely closed canopy or cover. The rainforest acts like a sponge, soaking up rainfall before releasing it slowly and steadily into the many rivers draining the areas. The plants also protect the soil from the heavy rainfall which would otherwise cause massive erosion.

The rainforest environment

The trees grow in multi-layered profusion: tall 'emergents' piercing the canopy; lianas, stranglers, and climbers with aerial roots festooning their buttressed trunks; lichens, mosses and algae adorning every surface; and an array of fungi colonising the forest floor. Almost every branch is hung with epiphytic* ferns, orchids, or bromeliads, while smaller trees and shrubs compete for light and space below. This intricate plant life supports an even greater diversity of insect and animal life, much of it specialised, with life cycles linked to certain plants.

Yet for all their intrinsic interest, these forests remain almost unknown to us. We now know more about certain sectors of the moon's surface than about the heartlands of Amazonia.

* growing on another plant.

Native peoples

Compared with other environments, people appear to have settled in the rainforests quite late. Earliest evidence of people living in the forests comes from South-East Asia and the Pacific Islands. Much later, but still some 20 000 years ago, it is thought that groups of people moved from Asia across the existing land-link to North America, down through the western part of the continent and into the Amazon rainforests of South America. Earliest traces of habitation in the African forests are no more

Kayapo Indians, of the Amazon rainforest, ready for a ceremony

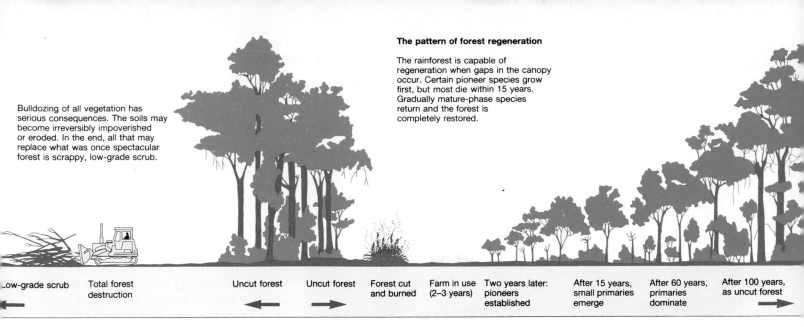

Bulldozing of all vegetation has serious consequences. The soils may become irreversibly impoverished or eroded. In the end, all that may replace what was once spectacular forest is scrappy, low-grade scrub.

The pattern of forest regeneration

The rainforest is capable of regeneration when gaps in the canopy occur. Certain pioneer species grow first, but most die within 15 years. Gradually mature-phase species return and the forest is completely restored.

| Low-grade scrub | Total forest destruction | Uncut forest | Uncut forest | Forest cut and burned | Farm in use (2–3 years) | Two years later: pioneers established | After 15 years, small primaries emerge | After 60 years, primaries dominate | After 100 years, as uncut forest |

The pattern of rainforest regeneration. A tropical rainforest can recover from being cut down, in time, but not if bulldozing has taken place

than 3000 years old – long after the building of the Pyramids in Egypt and Stonehenge in Britain. These earliest forest inhabitants were mostly hunters and gatherers of natural resources, which explains the lack of much evidence of their existence. Isolated from the rest of the world, these people evolved distinctive and stable ways of life, sometimes completely unaffected by events elsewhere.

In addition to hunting and gathering, many groups cleared patches of forest to cultivate a few crops and build temporary settlements. The soil is not naturally rich, so the community needed to move on after a few years to clear and cultivate another section of forest. Due to wise practices, evolved over thousands of years, the group would move on before the forest was permanently damaged, and it soon grew again. This system of shifting cultivation or 'slash-and-burn' clearing works provided the forest has enough time to regenerate. In the past little damage was caused but since the 1950s the pressure of population has become too great over wide areas: forests are not given time to regrow, the system leads to the destruction of both forest and soil, and the means of survival disappears.

Are the forest people 'undeveloped'? They have certainly acquired the skills to survive and flourish in conditions that would be impossible for most settled people. Over the centuries they have discovered a rich supply of foods and medicines in the plants and animals around them. Until recently, they were able to live in harmony with the forest environment, using and managing its resources without causing its destruction. On the other hand, they live without many of the material goods that most people today regard as a necessary part of being 'developed'. They also have a short life expectancy, often dying from diseases easily cured by known medicines, and although very skilled in survival they lack any formal education.

In all, a thousand or so tribes are scattered around the rainforests of the world. They differ considerably, as a comparison of the pygmies of Zaire and the Yanomani of the Amazon makes clear. In their uninterrupted state they had learned to live with their environment, and often had standards of living better than many settled peasant farmers. Much change has recently affected the rainforests, however, and most of the tribes are close to extinction or at best a dramatic change of life-style.

1 Explain the following terms: habitat, ecosystem, hunting and gathering, shifting cultivation.
2 Describe **a**) what you think would be the most difficult conditions to cope with if you unexpectedly found yourself living with an Amazonian rainforest group (apart from language), and **b**) what you think someone from the rainforest would find difficult to cope with in your environment (apart from language).
3 In what sense are the rainforest peoples **a**) more developed and **b**) less developed than people in your neighbourhood.

Exercises on page 205
World map of rainforest areas, and the rainforest climate

41

2.10 Clearance of the rainforests

Recent decades have seen an alarming destruction of rainforest environments. Between 1960 and 1985 the forest cover of Thailand was reduced from half to a quarter of the country, and the cover of Central America fell from 80 to 40 per cent of the land area. It is calculated that since 1945 about half of the world's rainforests have been cut down, and the present annual rate of loss is more than the area of England and Wales. The causes of this massive clearance are varied, but the consequences are causing world-wide concern.

Commercial lumbering

There is a large and increasing demand for the hardwoods found in rainforests. Strong and beautiful woods such as mahogany, ebony and teak are used in furniture, panelling, house and boat construction, and so on. Logging companies and governments in whose lands the trees grow obviously want to exploit this resource. Only the few mature trees within an area are needed, but logging operations cause considerable destruction to the surrounding forest. The construction of roads and logging camps adds to the damage. There is growing anxiety that new techniques will allow the production of wood-pulp and wood chips from hardwoods as cheaply as from the more rapidly growing softwood forests. At long last this total destruction of rainforests for commercial logging is being questioned, and pressure is being put on companies to at least follow planned programmes of cutting, reforestation or planting to allow the cut forest to be restored.

(Top) an area of the Amazon rainforest after being burned. (Left) a lumber road cutting through the West African forest

Fuelwood collection

It is not widely realised that much forest is lost by small-scale cutting for family needs. It is estimated that for some 250 million people the only source of fuel is the wood they collect, and much of this is taken from the rainforests and their margins. The wood is needed for cooking and keeping warm, and until some alternative is found for the poorer peoples of the world they will understandably continue to destroy these environments.

Clearance for farming

Forests have always been cleared for farming since men and women learned the methods of cultivation. We have seen how the indigenous forest people still practice the slash-and-burn methods used for thousands of years by their ancestors. What is new is the scale of the operation, and the failure of the cleared forest to replace itself.

Some of the cleared land is used for the commercial production of crops suited by the climate. Large plantations and estates have been carved from the forest, usually with capital from economically developed countries elsewhere. One reason why such countries took these lands as part of their empires was to obtain supplies of resources and raw materials such as rubber, cocoa, palm-oil and so on. More recently huge tracts of forests have been cleared to provide grazing land for cattle. Once cleared the land is so poor the ranches provide little more than a few years' pasture. The land is so cheap, however, that it pays to produce beef from these areas, particularly in Central and South America, for the hamburger and fast-food companies of North America.

Very different from these are clearances for new settlements. Sometimes governments wish to encourage massive migrations from overcrowded parts of their countries where population pressure is high and land redistribution is not a possible or acceptable answer. Examples of this can be seen in movements from Indonesian Java to clearances in Borneo, and in Brazil with

Once the rainforest has been cleared, commercial planting begins

movements from the barren savanna lands to rainforest clearances. Apart from this, thousands of peasant farmers migrate of their own accord to colonise the forest margins in an attempt to make a better life.

Consequences of clearance

Unfortunately, the areas cleared of forest provide a poor farming environment. Not only are the soils naturally lacking in minerals, but once exposed to the heavy rain they get washed away. The native peoples, as elsewhere in the world, are ruthlessly moved as their traditional habitats are destroyed. Last but not least there is a real fear that removal of forest cover on this scale will cause changes in the world's climate affecting us all.

1 For the new uses of forests shown in these photographs explain who is benefiting from the activity, and how. Who is losing, and how?
2 Argue the case for *or* against clearing and using land inhabited by traditional forest people.

Exercises on page 206
Conservation of the rainforests

43

2.11 Environmental disasters

As we have seen, some environments are harsh and extremely challenging to human survival. Peoples who have lived in the arid, polar, rainforest and mountain lands for thousands of years have had to adopt ways of life suited to the severe climates and habitats. Until recently these peoples have been relatively few in number and very isolated from the rest of the world. Although their material standards of living may seem lower and less developed than elsewhere in the world they have shown remarkable skill and determination in living in such conditions and regarding them as their homelands.

In contrast to these places are those which can normally support people in considerable numbers but are liable to experience environmental disasters. Shattering events such as earthquakes, volcanic eruptions and cyclones with accompanying storm surges can cause great damage and loss of life in otherwise productive and settled areas. Bangladesh is one of the poorest countries of the world, and it is sometimes claimed that this is partly due to the environmental disasters that hit the delta lands.

Cyclones and storm surges in southern Bangladesh

Much of Bangladesh consists of the delta created by the Ganges and Brahmaputra and their tributaries that drain the Himalayas to the north and flow into the Bay of Bengal. This remarkably flat flood plain is one of the largest in the world. It experiences torrential monsoonal rains between June and September as saturated air masses sweep in from the Bay of Bengal. The Ganges and its tributaries regularly flood the low-lying land, leaving a layer of rich silt over the fields. Temperatures are high, and providing the rains fall, up to three crops of rice can be grown in a year, as well as the commercial jute crop. Many cattle and other livestock are kept in villages and small towns scattered over the delta. About 25 million of the country's 90 million total population live in the delta region.

In some years the monsoon rains fail to arrive and there may be drought and shortage of food. Equally likely, there may be severe flooding from the many rivers threading the delta, with damage to crops, buildings, power and water supplies and road and rail links. The greatest devastation, however, comes when the area is hit by a tropical cyclone moving in from the Bay of Bengal. Torrential rain falls from towering clouds and winds of enormous force blow in over the coastal sea and the delta. Huge waves at sea are a threat to shipping and fishing boats. Most destructive of all, the storm surge and hurricane-force winds sweep over the coastal communities, drowning people and livestock, washing away crops and damaging all buildings in the path of the storm.

About 50 or 60 tropical cyclones develop over the world's seas every year, the majority never reaching land. But when they do, the results can be devastating. Over the past 200 years the delta lands of what is now Bangladesh have experienced an average of one

Harvester trying to save some of the jute crop after flooding in Bangladesh. Jute is the main source of export earnings. Rice is the major food crop, and it too can be destroyed by floods

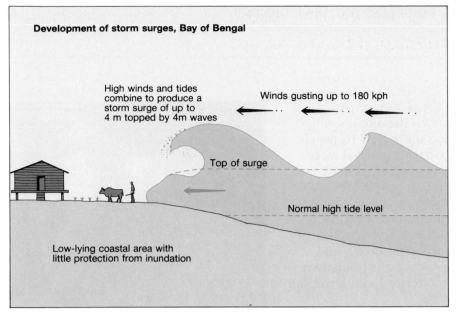

Development of storm surges, Bay of Bengal

High winds and tides combine to produce a storm surge of up to 4 m topped by 4m waves

Winds gusting up to 180 kph

Top of surge

Normal high tide level

Low-lying coastal area with little protection from inundation

This diagram gives a good impression of the frightening scale of a storm surge

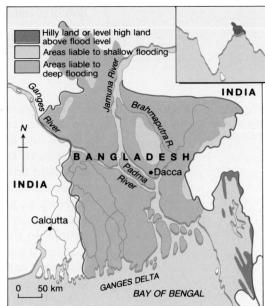

Hilly land or level high land above flood level
Areas liable to shallow flooding
Areas liable to deep flooding

INDIA

Ganges River

Jamuna River

Brahmaputra R.

BANGLADESH

Padma River

•Dacca

INDIA

Calcutta

0   50 km

GANGES DELTA

BAY OF BENGAL

Bangladesh and the Ganges Delta

cyclone disaster every five or six years, although the frequency is very variable and cyclone occurrence unpredictable. The most dramatic ever recorded occurred on 12th November 1970 when over 300 000 people perished, hundreds of thousands of cattle drowned, 400 000 houses were damaged or destroyed and well over half the fishing fleet was lost. In May 1985 a storm surge drowned an estimated 11 000 people soon after the area was trying to recover from disastrous river floods.

The combination of low land, a large tidal range, the funnelling shape of the coastlines focusing on the delta, and the possibility of severe cyclones and storm surges coinciding with high tides, makes this a precarious place in which to live. Yet 25 million people continue to do so in spite of such recent massive loss of life. They are mostly peasant farmers or landless labourers. For any region to support a large population at a high density it is necessary for the advantages to outweigh the disadvantages. In the delta lands of Bangladesh these are finely balanced. The government lacks money to provide an effective early warning system or means of rapid evacuation if disaster threatens. The people are too poor to move elsewhere and have to accept the risks, hoping that, as often happens, the cyclones will veer away from the region. The regular and terrible environmental disasters do

not cause the low standard of living of the delta people. Rather, the low standard of living dictates that the people have little choice other than to accept their fate and that the government has few means to provide any protection.

1   Suggest why 25 million Bangladeshis are living in an area which is quite likely to experience devastating tropical cyclones.
2   In 1985 there were two major disasters: an earthquake destroyed much of the centre of Mexico City, and a mud-flow resulting from the eruption of Nevado del Ruiz destroyed the town of Armero in Colombia. In both cases many people lost their lives. Compare the causes and consequences of these environmental disasters with the cyclones and storm surges.

Exercises on page 207
Tropical storms

Hurricane damage on the Florida coast, USA. Environmental disasters can affect even wealthy areas

2.12 Disease

(Right) stagnant pools like this are a breeding ground for mosquitoes

One measure of the wealth or poverty of an area or country is the amount of illness and disease suffered by its people. Illness can range from a general weakness and fatigue to crippling disease and premature death. Sick people are unable to work as productively or contribute as much to the community as healthy people. In that sense, if an environment leads to a great deal of ill-health and disease it can be said to contribute to a lack of economic activity and a lowering in the quality of life.

Water-related diseases: river blindness and malaria

Many of the most destructive of human diseases are related to water and water supply. In some cases the water acts as the breeding ground for insects which then carry the particular disease from infected to uninfected people. It is estimated that more than 20 million people who live in tropical parts of Africa and Latin America suffer from onchocerciasis or so-called river blindness. River blindness doesn't kill, but

Children lead adults blind through onchocerciasis (river blindness) to work in the fields, Burkina Faso, West Africa

often leads to impaired vision and total blindness. It is caused by tiny worms that infect the skin and finally scar and damage the eyes. These tiny worms enter the skin when a person is bitten by a small black fly that has previously bitten an infected person. Not only does the disease cause great personal suffering, it also means that infected men and women are an additional burden on the community. Whole villages may be deserted if widespread infection occurs.

In some of the savanna lands of West Africa, such as the zone near the River Volta in Ghana and the River Hawal in north-east Nigeria, a high percentage of men are affected. The flies that carry the disease breed in fast-flowing rivers near rapids where there is a lot of oxygen in the water. Local people have to use the river for water in the absence of supplies from deep wells, and are vulnerable to infection. It is difficult to eliminate the flies completely, though there has been increased control in the Volta area. Drugs have not yet been developed to cure the infection. So millions of people are doomed to poor vision and the probability of blindness at an early age.

Malaria is another disease carried from person to person by insects – this time by mosquitoes. It is estimated that every year about 800 million people suffer from the fever, and many die. The disease is most widespread in the hot wet parts of the world, and in Africa south of the Sahara malaria kills one in two children under the age of two years who become infected. Malaria is likely to be a threat wherever mosquitoes breed – in stagnant pools, in lakes, in pots of water and so on. One solution that has proved effective is the spraying of the breeding sites with DDT and other insecticides (but these can pollute the environment). The mass use of drugs has also had a big impact on preventing or limiting malaria, but as the mosquitoes become resistant to the various insecticides and the malaria parasites infecting humans become resistant to drugs, the disease is re-emerging and remains a major cause of misery and death.

Water-borne diseases: bilharzia and Guinea worm

In contrast to these insect-carried infections some diseases are caused through direct skin contact with infected water. A person infected with bilharzia contaminates a pool or stream by passing eggs into the water, and the tiny larvae which hatch live first in water snails and then swim freely as tiny worms. These can be picked up by anyone standing, walking or swimming in the water. The infected people feel weak and ill and unable to work. Dams and irrigation schemes that should lead to improvements in living standards unfortunately often provide the weed-covered lakes and canals in which the water snails breed.

Guinea worm is caught by drinking water that has been infected by someone who already has the disease. The worms grow to 30 mm or more in size before breaking through the skin, usually in the feet or legs below the knee. Tiny worms from these wounds enter the water and are kept alive in very small creatures known as cyclops. It is these that are taken in when drinking contaminated water. Millions are affected during the peak season of infection, and many are made weak and ill year after year. The simple cure would be for infected people not to contaminate the

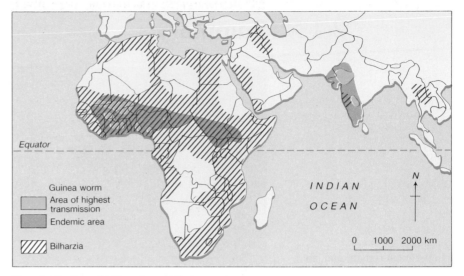

The distribution of bilharzia and Guinea worm

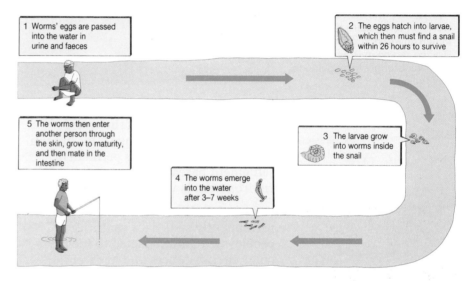

The life cycle of the bilharzia worm

water, and for uninfected people to drink water only from deep bore-holes or protected wells, and not to step in streams and pools. But that is not easy to do. There are no effective drugs to cure the disease, but it could be eradicated with care and the provision of clean water supplies.

1 Suggest why **a)** breeding grounds for mosquitoes and the black fly that causes river blindness are not cleared or destroyed, **b)** people in areas of high infection continue to stand in and drink contaminated water, **c)** drugs are not available to cure river blindness and Guinea worm, and **d)** few people in countries such as the UK suffer from water-related diseases.
2 Describe the distribution of Guinea worm, naming six countries with high transmission.

A Guinea worm which has erupted through the skin

Exercises on page 208
Combating malaria

47

2.13 Summary

In the most remote and least accessible parts of the Third World, air transport might be the only, and is certainly the fastest, link with the rest of the world. In this picture Kayapo Indian children from the Rio Branco region in Brazil carry supplies brought in by plane to their village

☆ Landforms, climates, plant cover and wildlife in some parts of the world provide harsh environments for human beings. In extreme cases permanent settlement based on the resources of the environment is virtually impossible.

☆ Some harsh environments have been permanently settled for centuries by peoples who have adapted to the particular circumstances and live from the local resources. Although at the limits of human adaptability they have lived in harmony with their environments.

☆ These indigenous peoples show great skill and endurance in surviving under difficult conditions, although compared with most people in more widely settled parts of the world – but by no means all – they have a low material standard of living and level of comfort. This is not to say that they regard their quality of life as inferior, though outsiders might.

☆ In some parts of the world poverty forces people to destroy the very environment on which they depend – by overgrazing and over-farming for food, and over-cutting woodlands for fuel.

☆ Development aid has sometimes added to the environmental problems of soil erosion, pollution, desertification and tropical deforestation.

☆ The main hope is 'sustainable development', in which present needs are met while trying to ensure that resources and environments will be available for future generations.

☆ Harsh environments may be the location of enormous resources, particularly minerals, that could never be exploited by the indigenous or local people. They can be exploited and removed with the vast wealth and modern technology of the industrialised countries of the world.

☆ These harsh and remote environments may also be regarded as important by outsiders for military, scientific or tourist reasons.

☆ The development of resources, siting of military and scientific bases and the growth of tourism may lead to economic material gain for the indigenous people – in the form of money for land, employment, or the provision of many amenities and goods.

☆ Impact on traditional ways of life may be dramatic, sometimes destroying them altogether. Alternative life-styles may not be available, or if they are, may seem inferior to what has been lost.

Chapter 3

Hundreds of thousands of people in Revolution Square, Havana, Cuba, celebrating twenty-five years of communism in 1984

People and development

To what extent are contrasts in development due to the characteristics of people within countries, and to the ways they organise their lives?

Population density and living standards

It is sometimes claimed that a major cause of low living standards in various countries is over-population, and that if people had fewer children their standard of living would rise. At first this seems a reasonable suggestion. People use the resources available within their environment for food and drink, shelter and clothing, and to provide the energy and goods that can improve their quality of life. The more people there are to share these resources the more thinly they will be spread. If the population increases at a faster rate than the development of resources then people will get poorer and poorer. There are, however, several errors in this argument, and amongst these is one related to the way the wealth of a country is actually produced.

High density of population and poverty

There is no question that extreme poverty and degrading living standards are found where people are crowded together in a small area. When people are directly dependent on the land and the food it produces there is a clear link between numbers, the size and productivity of the land, wealth produced and living standards. Densely settled farming areas where people live from what they produce may be prosperous, but are often poor. If they are poor it is often because the land and produce is owned by a few relatively rich villagers and the wealth is not available for everyone. Families with only small patches of land, or those of landless labourers, find it hard to survive. On the other hand the problem may not be of ownership, but that there are too many people for the land to adequately support. Only in this latter case can it be said that low living standards are due to over-population.

One solution to rural over-population has been to migrate to cities to find work, but all too often the situation in the cities is worse rather than better. Millions of people live in cramped and overcrowded shelters with inadequate amenities for a decent standard of living. These may be in inner city slum areas or, more commonly in Third World cities, in large squatter settlements on the outer fringes. Many survive from

World distribution of population, and statistics for six selected countries

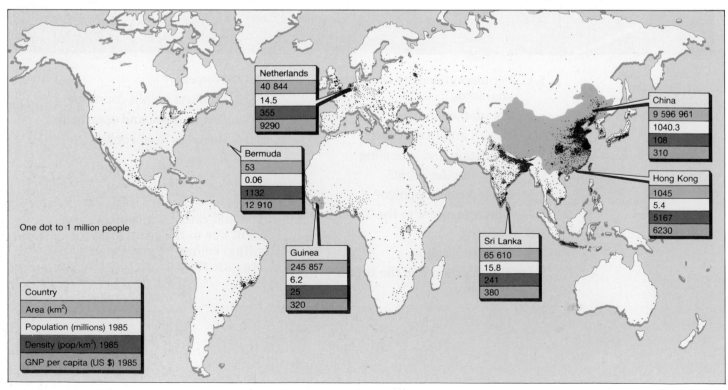

Netherlands
| 40 844 |
| 14.5 |
| 355 |
| 9290 |

China
| 9 596 961 |
| 1040.3 |
| 108 |
| 310 |

Bermuda
| 53 |
| 0.06 |
| 1132 |
| 12 910 |

Hong Kong
| 1045 |
| 5.4 |
| 5167 |
| 6230 |

Guinea
| 245 857 |
| 6.2 |
| 25 |
| 320 |

Sri Lanka
| 65 610 |
| 15.8 |
| 241 |
| 380 |

One dot to 1 million people

| Country |
| Area (km²) |
| Population (millions) 1985 |
| Density (pop/km²) 1985 |
| GNP per capita (US $) 1985 |

part-time and casual work, and even those in full employment often get very low wages. With so much poverty in high density rural and urban areas it is not surprising that over-population is blamed for low living standards. But it should not be forgotten that some of the poorest people in the world live in small, scattered, low-density communities.

High density of population and wealth

Some of the wealthiest and most advantaged people live in high-density urban areas, although rarely with anything like the number of people per room or dwelling as found in the poorest areas. In some countries millions of families of modest means lead reasonable lives in huge, densely packed cities. In some favoured areas, where the environment is suitable and wealth and technology available, there can also be quite dense and prosperous rural communities.

The explanation for such wealth in high-density areas is that such people do not depend on the limited produce of the land on which they live. They are not subsistence farmers, but live by trade or providing goods and services to others. In these circumstances there is no longer a simple relationship between numbers and area of land. Through using their skill and experience, resources, acquired wealth or control over others they can buy the food and other things they need. Space is not the great need. In this way it is possible for there to be a high density of population and relatively high living standards. Two very different places that illustrate this combination are the Netherlands and Hong Kong – though it should be remembered that both contain considerable differences. Neither could be considered 'over-populated'. The wider world is the source of their wealth, as well as the land which they occupy.

Even in countries where there is a more direct link between people and production from their land it is dangerous to explain poverty through over-population. It may be that the land could produce quite enough if sufficient money and proper technology were available, or if wealth from the land was more evenly distributed amongst the

population. There are many ways in which production and living standards could be improved without reducing the population. Nevertheless, population numbers are a part of the equation, and as we shall see, the growth of population is a concern of individual families, nation states and the whole human race.

1 Look at the map, and name the major areas of high population density.
2 Using the figures in the tables on the map to illustrate your answer, discuss whether there is any relationship or correlation between population density and GNP per capita.
3 For one of the three countries with a high GNP per capita describe its location and features from the map and statistics and suggest how it manages to support a high density of population.

High density, poor quality housing in Hong Kong. In other parts of Hong Kong there is high density, good quality housing

Exercises on page 209
Photographs relating to the theme 'population density and living standards'

A Netherlands landscape – another type of 'greenhouse effect', and an example of high density of population and wealth in a rural area

3.2 Population change

The statistics shown here are the results of a survey in three different countries to find out why people have children

MEXICO

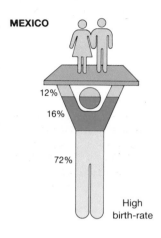

12%
16%
72%

High birth-rate

SINGAPORE

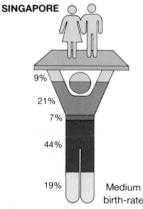

9%
21%
7%
44%
19%

Medium birth-rate

USA

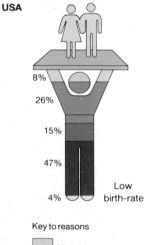

8%
26%
15%
47%
4%

Low birth-rate

Key to reasons

Various

Psychological benefits to parents

To strengthen the marriage

Companionship

Economic support

The total population of the world is increasing at a rate that is now recognised as potentially disastrous. In the late 1980s it had reached about 4800 million, and it is expected to rise to about 6000 million by the year 2000. The 75 million or so 'extra' people added to the world's population each year in the late 1980s will be nearer 90 million a year by the turn of the century. Much depends on what happens to birth-rates and death-rates, but it is calculated that the world population will be between 10 and 14 billion in a hundred years' time. Pressure on global resources will be two to three times as great as now, and that is a cause of great concern.

To achieve a rough stability in the balance of births and deaths, the average family should consist of two children. The average family in Europe and North America has less than two, in Asia and Latin America more than four and in Africa more than six. These are average figures and mask great differences between individual families. The main point is that population increase is greatest, and likely to remain greatest, in countries experiencing least economic growth and with lower than average standards of living. It should not be

forgotten, however, that low rates of increase, or declining numbers, produce an 'ageing' population. This brings its own set of problems for a country.

Population change in a country is due either to natural causes – births and deaths – or migration of people into and out of the country. In-migration is sometimes a sign that there is plenty of employment available, usually in new industries or building and construction work. All too often, however, migrants are leaving an area of natural disaster, or are refugees from political persecution. In many cases immigration and emigration occur at the same time and there may be little overall effect on the total population. A great deal of migration for work is also temporary, and doesn't affect the population totals.

A decline in the death-rate is usually the result of improved health care and living standards (apart from the ending of a war). Birth- and death-rates are usually given as percentages or as a number per thousand of total population. From the figures given here it can be seen that the death-rate in many parts of the world has fallen very considerably. Birth-rates are also falling in most places, but they still differ greatly

A mother and her five children. Mali, West Africa

from country to country. The average in Africa is about 46 per thousand, while in Europe it is nearer 14 per thousand. It is noticeable that birth-rates are often very high in countries that also have a high infant mortality rate.

Population planning

The rate of change of population is closely related to changing birth- and death-rates. Most would regard a reduction in the death-rate as desirable, especially if it meant a drop in the needless deaths of infants and young children. But it would lead to dramatic population growth. There is great controversy over control of birth-rates, especially if it is by the use of contraception devices. Some people choose to limit the size of their families, and many governments encourage this through advertising, financial incentives or penalties, and help with programmes for family planning or sterilisation. On the other hand many individuals and groups oppose both birth control and abortion on religious and moral grounds. In countries where children are regarded as an economic asset, security in old age, or a sign of status for men, then reduction in the size of families is again not seen as desirable by everyone. Whatever the national policy, the acceptance or rejection of birth control measures by individuals is linked to many factors such as education level, material wealth, religious beliefs and social customs.

Variations in rates of population change

We have seen that people are distributed very unevenly around the world, with great variations in density from place to place. There are also marked variations in rates of population change (usually population growth) between countries. Towards the end of the 1980s about three-quarters of the world's population lived in the Third World, and this proportion will rise to about 80 per cent by the year 2000. Some countries at present classified as 'developing' may by then have made great progress in economic development and improvement of living standards and so might be included with the more developed nations. But all the signs are that during

Population statistics for selected countries

Country	Birth rate per thousand 1965	1985	Death rate per thousand 1965	1985	Average annual growth of population (per cent) 1980–85	Total population (millions) 1985
USA	22	16	9	9	1.0	239.3
Mexico	44	33	11	7	3.2	78.8
Kenya	51	54	21	13	4.1	20.4
Kuwait	48	34	7	3	4.5	1.7
India	45	33	20	12	2.2	765.1
China	39	18	13	7	1.2	1040.3

the twenty-first century most of the world's population growth will be in the present-day developing countries.

1 a) Compare the birth-rate in 1985 and the change over the previous twenty years for Kenya, India and the USA. Explain the differences. b) Compare the change in death-rates over the twenty-year period between Kenya, India and China. Explain the differences.
2 Look at the diagram, 'Why people have children'.
a) How do the main reasons for having children vary between the three countries?
b) Why, or in what way, is the information shown by the diagram flawed?
3 Argue the case for or against the statement 'Eradication of poverty is the best form of birth control'.

> **Exercises on page 210**
> Population statistics, and world map showing rates of population increase

A major family planning campaign in China has helped to bring down the rate of population increase

3.3 Health and health care

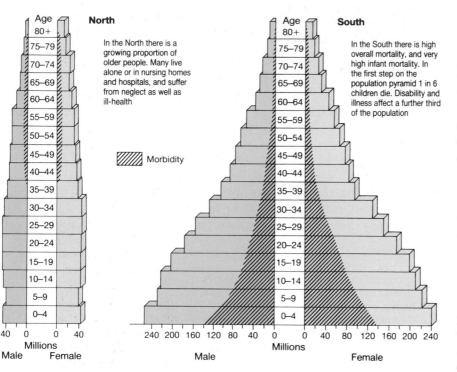

Age | North

In the North there is a growing proportion of older people. Many live alone or in nursing homes and hospitals, and suffer from neglect as well as ill-health

Morbidity

Age | South

In the South there is high overall mortality, and very high infant mortality. In the first step on the population pyramid 1 in 6 children die. Disability and illness affect a further third of the population

Population pyramids for North and South (pyramids for individual countries would show a similar structure). Morbidity is the spread of disease within the population

For world map showing life expectancy at birth, see page 115

It is very difficult for anyone who is ill or permanently undernourished to work efficiently or get much satisfaction from life. Ill-health and undernourishment can affect people of all ages and in every part of the world, although the type, distribution and impact of disease differs a great deal between the North and the South. We have already seen that hundreds of millions of people in the developing world suffer from water-borne and other environmental diseases such as bilharzia and river blindness for which no vaccine exists. Vast numbers are also made ill or die from malaria, cholera, typhoid and many other infectious diseases. Illnesses resulting from undernourishment and malnutrition (a lack of proteins and vitamins) affect many millions more, particularly young children and pregnant women.

These illnesses make people extremely ill and weak, and unable to do any sustained work. In developing countries even a short period of ill-health can push a family into a severe crisis. Unable to earn a wage or produce food a family may have to sell what little

land or few animals or other possessions it has, thus adding to its poverty. Yet it is poverty that is a major cause of the malnutrition, poor sanitation, lack of water and inability to obtain health care that leads to ill-health in the first place. Widespread ill-health is a significant contributor to the lack of economic development in many Third World countries.

Infant and child deaths in developing countries

The population pyramids show in a very general way the different population structures and morbidity patterns of the North and the South. Patterns of mortality are also different. Death-rates are much higher in the Third World countries, and this is particularly marked among young children. Over 95 per cent of infant deaths, that is between birth and one year of age, are in countries of the South. The average infant mortality rate is about 100 per thousand births, compared with an average of 19 per thousand in the North. In some Third World countries this rate is far higher. The main causes of infant deaths are infectious diseases, malnutrition and, the greatest killer of all, diarrhoea and its effects. Measles, whooping cough, polio, tuberculosis, tetanus, diphtheria and so on could be controlled by immunisation, as they largely are in countries of the North. But few children in the Third World are protected in this way, mainly because effective vaccination depends on cold storage of unstable vaccines and the giving of regular boosters. The major cure for diarrhoea is oral rehydration therapy (ORT), which consists of giving patients a simple mixture of sugar and salts in water. The introduction of ORT has had a big influence on controlling diarrhoea deaths in infants, but education in its use is still a major task.

These Indian children grow up on the street. Their health prospects are poor

Health care in the South

During the latter part of the twentieth century the main causes of ill-health and death in the countries of the North are heart disease and cancer, accidents on roads and illnesses related to industrial activity. Increasingly there are fears of widespread disease or death through nuclear accident or warfare. In the South the problems of ill-health and preventable death are more closely related to malnutrition and infectious disease, both of which could be controlled if money and resources were available.

Many people argue that it is far better to use the limited resources available to countries in the South to improve health care at the village and local level rather than in large and expensive modern hospitals. This approach, sometimes known as 'primary health care', consists of training health workers to give basic advice to people in scattered rural communities on hygiene, healthy eating, drinking clean water and improving sanitation. Help and advice is also given on family planning, and dispensing simple drugs. Sometimes this advice is in the face of both traditional practices and customs and the persuasive methods of firms selling drugs or items such as milk-powders for babies.

Trained nurses and teams of volunteers support the efforts of the village health workers by visiting scattered clinics and undertaking immunisation programmes and health surveys, while aid teams try and help with practical tasks such as providing deep wells and sanitation schemes. These in turn are supported by a trained doctor at a regional hospital in a nearby small town. Only the largest towns and cities have the expensively equipped hospitals capable of undertaking major operations. In this way primary health care is made available to many millions of people, and this has an overall effect of improving the general health of large numbers of the population. Health care for all, with a minimum of preventable ill-health and death, is a basic measure of development within a country.

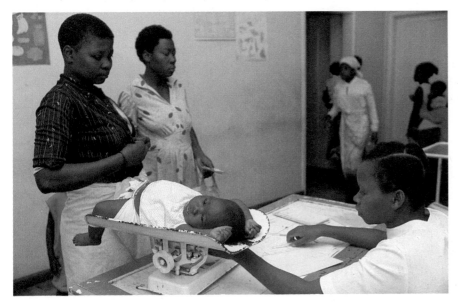

Primary health care in an African clinic

1 Why is ill-health in a family much more disastrous to the average family in a Third World country than in say, the UK or Australia?
2 Compare the population structures and morbidity patterns of the North and South. What effect does a large number of very old or very young people have on the economy and standards of living of a country?
3 Compare the patterns of mortality of countries of the North and the South. Try to explain the differences.

> **Exercises on page 211**
> The Big Six diseases and immunisation

3.4 Education and schooling

Most countries spend a considerable proportion of their wealth on schools and the education of their young people. In many of them, in addition to this state provision, parents will also pay for their children to attend state or private schools. Education is clearly regarded as important for both the individual pupil and the future of society. But as with most things the provision of education is unevenly distributed around the world.

Literacy around the world

It is easy to understand why migrants from a country with a different language find it hard to read or write when they first arrive, but more difficult to accept that in some parts of the world a majority of adults cannot read or write the language of the country in which they were born and grew up. Levels of illiteracy are falling in most countries but there are still almost 850 million illiterate men and women in the world. They

face enormous disadvantages. They cannot read newspapers or books, letters from friends or official notices. They cannot write to their family or friends. The acquisition of useful knowledge, understanding and skills is made more difficult. Most illiterate people are very vulnerable and dependent on others for information, instruction and to communicate with others.

The proportion of illiteracy varies a great deal from place to place. It is greatest in Africa where, in spite of great efforts to improve the situation, 74 per cent of the people are illiterate. In Asia the average is about 47 per cent and in Latin America just under 25 per cent. Literacy levels differ greatly for men and women in most countries, reflecting different social attitudes to the role of and respect for women.

The illiterate person is not only unable to read or write, but is more likely to be poor, unemployed, hungry and vulnerable to disease. A global map of illiteracy closely matches those for poverty, malnutrition and ill-health. Illiteracy represents a great underuse of human resources as well as personal disadvantage.

Inside a Nepalese village school

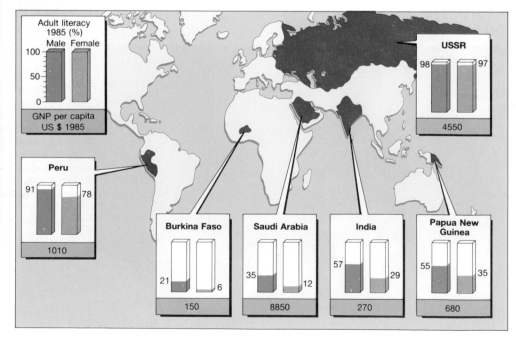

Adult literacy rates and GNP per capita in five Third World countries, with the USSR for comparison

Schools and schooling

Although formal schooling is widely regarded as a vital means of helping young people become literate and more widely educated, the proportion of GNP given to schooling has been dropping in most countries. It is calculated that in the developing countries at least 120 million children do not have a school to attend. When they are available there is a high rate of drop-out. Many who start elementary school attend for one year or less, and only about 60 per cent attend a four-year course. Reasons include the need for the children to work to contribute to the meagre family income, ill-health and, in many cases, the high costs to parents.

There is also criticism of much formal schooling in developing countries. School buildings are often little more than a roof and four walls, and teaching resources are frequently in desperately short supply. Teachers are dedicated and caring, but often have had only a limited education themselves. The work is often formal and based on past European practices from the days of colonial rule, rather than on the needs of the community and country today.

Much is geared to preparing a small proportion for secondary and higher education. Only a tiny minority benefit from college and university education and become the technicians, scientists, doctors, administrators, farmers and so on that their countries need.

It seems that conventional schooling cannot solve the problem of illiteracy in the developing world, nor provide the general education for personal satisfaction and the needs of society. In that sense lack of education contributes to lack of economic and social development. But all the evidence is that people in developing countries are perfectly capable of attaining full literacy and the highest standards of education and training. It is poverty that causes low educational attainment rather than the other way round.

This Muslim boy from Northern Nigeria learns to recognise arabic script so that he can recite the Koran

A practical gardening class in Papua New Guinea

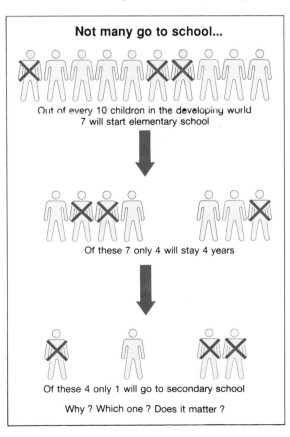

Not many go to school...

Out of every 10 children in the developing world 7 will start elementary school

Of these 7 only 4 will stay 4 years

Of these 4 only 1 will go to secondary school

Why ? Which one ? Does it matter ?

1 Describe and explain the differences in adult literacy for the countries shown on the map.
2 Compare the features of the Nepalese 'school-room' with your own. What surprises you about any similarities or differences?
3 What are some of the reasons for the drop in the share of GNP's allocated to education? What priority do you think education should get?

In the Third World, only one child in ten, on average, gets a secondary school education

Exercises on page 212
Education statistics

3.5 Education for what?

An examination in progress in Pakistan

Primary education is a neglected area in the Third World. Education might also be more wide-ranging

In the Third World secondary and higher education for a few receives most of the money. Perhaps more priority should be given to providing all children with a basic education. Such an education would help them control and improve their own lives, and might include the things shown in this diagram.

Literacy

Child-care and home running

Skills related to employment and income opportunities

ONLY 6% OF FOREIGN AID FOR EDUCATION IS ALLOCATED FOR PRIMARY SCHOOLS

Methods of increasing land productivity

Environmental understanding

Nutrition, health, hygiene training

House improvement, construction skills

Education for participation in community and political life

Numeracy

Important though literacy is to the quality of life of most people, it is only one of the skills that can be developed in schools. Some other valuable contributions that might be made are suggested in the diagram. Unfortunately they are rarely provided, and when they are it is frequently with limited success. There are many reasons for this, but in many Third World countries a major cause is the importance of passing examinations in getting a job and some influence in local affairs. The surest way of doing this, because of the traditional style of examinations inherited from the days of colonial rule, is through rote learning of information and the mechanical practice of language and mathematical skills. As we have seen, many pupils leave elementary school after one or two years, while only the more able remain to sit examinations. From these a select few go on to secondary education and then to college or university to take yet more examinations.

Quite obviously a number of very able men and women are trained for important jobs in this way, and both they as individuals and the country as a whole benefits from their education. But millions experience a schooling that is not very helpful to them. It would be more useful if as well as becoming literate and numerate they could also develop some of the practical skills mentioned in the diagram, as well as learn something about their local community and the wider world. Apart from the domination of formal teaching for examinations, though, there are other difficulties. There is often a lack of basic resources for learning and teaching, and schools and equipment may be very simple. Teachers may themselves lack training in alternative methods, or due to financial needs be forced to take other jobs as well as teaching. In many places political or religious leaders who control schooling may see them as places for religious or political indoctrination, and not want thinking, critical, well-trained and self-reliant young people.

Ethiopian women undergoing training in basic literacy (above) and electronics (above left)

Alternatives to schooling

It is increasingly believed that the difficulties are so great that there is little hope of formal schooling meeting the needs of millions of children – and adults – for education and training. Many alternative methods are being tried, and great hope is placed on radio and television for 'distance learning'. Remote parts of the world can have access to quality educational programmes and skilled teaching through the use of satellites and receiving dishes. Film, music and drama can also be used to support the formal teaching in schools. In many parts of the world youth groups, adult education centres, women's organisations, co-operatives and workplaces and village councils provide important opportunities for learning and training. But these new methods and centres need money just as schooling does, and that is what is lacking in Third World countries.

More than anything else, motivation matters in education and training. Wanting to learn is as important as good resources and skilled teachers. A major breakthrough in adult literacy in Latin America was based on encouraging reading by relating it to real practical needs and issues of everyday life. It is obvious from the remarkable achievements of many educated and skilled men and women in the Third World countries, as well as from the success of adult illiterates in learning to read and write when properly motivated and supported, that it is quite wrong to talk of their lack of ability. Economic and social develop-

ment of a country may well be handicapped by the lack of enough engineers, agricultural scientists, doctors, teachers, scientists, technicians and so on. That shortage is not due to the lack of ability of its people, but a lack of money and resources to provide the education and training that they need and could benefit from. It is a vicious circle that is hard to break.

1 Draw a diagram to represent the 'vicious circle' between underdevelopment, lack of money for education and training, and lack of trained and educated people to enable developments to occur.
2 Which languages and educational systems are most likely to influence schooling in Nicaragua, India, Jamaica, Senegal, Indonesia?
3 **a**) Why should the action of the police in Bangladesh so anger the students? **b**) Why should some religious and political leaders not want the sort of activities described in the diagram in their schools?

Exercises on page 213
Learning in an African village

Exam clashes lead to Bangladesh curfew

A curfew was clamped on the Bangladesh town of Magura after three people were killed and more than 55 injured in clashes between police and students on Thursday, officials said yesterday.

The officials in Magura, about 150 miles west of Dhaka, said two of those killed in Thursday's clashes were students. Two policemen were still missing after the violence, during which police fired more than 100 rounds and used batons to disperse an angry crowd.

The town's chief administrator said the clashes started when police tried to stop people supplying answers to students during a graduation examination. A crowd attacked the police, burned government vehicles and ransacked the house of the police chief.

3.6 Skills for development

It is sometimes suggested that the main reason living standards are low in Third World countries is because their people lack the practical skills to do the jobs that create wealth. For example, it is claimed that less food is produced than might be because of the lack of under-standing and use of efficient farming methods, and that when plantations, mines or oil wells, factories or construc-tion projects are started foreign experts have to be employed because local people can't do the work. These sugges-tions are often true, but the important question is whether local people could do these jobs if given training and experience.

Two fine examples of traditional skills from Indonesia; rice terraces (below) and weaving (right)

Looking at the past

Even a brief glance at the past of Third World countries reveals a wide range of remarkable achievements. The Cheops Pyramid and the nearby Sphinx are just two of the incredible constructions built along the banks of the River Nile in Egypt more than 4000 years ago. Even today there is uncertainty about the methods used, and undoubtedly a great deal of slave labour was used to haul the massive stones into position. But there is no question that these and other pyr-amids show that the people of north-east Africa possessed great design, construc-tion and technological skills. Over the following centuries similar and equally impressive building took place in areas as scattered as present-day Mexico, Peru, Zimbabwe, Thailand and China. Road networks were built equal to anything Europeans could construct. Metals and precious stones were worked to produce beautifully crafted objects in the villages and towns of the Andean mountains and the rainforests of West Africa. Great terraces were built across steep hillsides so that crops could be grown. So well were some made that they still remain intact, several thousand years later. It is more difficult for other sorts of evidence to survive, but there is enough to show that people who lived in what are now the Third World lands had great knowledge of weather and soils and practiced very efficient farming methods. They were ingenious inventors of tools and equipment, clever mathematicians and astronomers, and were able to perform remarkable surgical operations. All in all it is an impressive record of great and varied talents.

Work skills in the modern world

The Industrial Revolution, with its new technology, use of machinery, new forms of energy and concentration of production in large factories, began in England and spread into Western Europe. Since then industrial methods and the ways of life that go with them have been widely adopted in North America, Australia and New Zealand, the USSR and Japan. Smaller concentrations of industrialisation are now found in many other countries, of which South Africa, Brazil, Korea and China are good examples. Meanwhile a second industrial 'revolution' is occurring with the use of electronics, automation and computers in economic production. Farming is also being revolutionised with the application of biological, chemical and technological discoveries. All these changes are affecting not only the production of food and goods, but also patterns of work and the opportunities for employment in every country of the world.

It is often helpful to divide economic activity into three broad groups. The Primary group covers activities such as farming, fishing and the mining or quarrying of raw materials. The Secondary group includes manufacturing activities. The Tertiary group, sometimes sub-divided into smaller sections, includes activities as varied as building and construction, transport, shops and other services, professional jobs and so on. The proportion of the workforce employed in these different groups varies

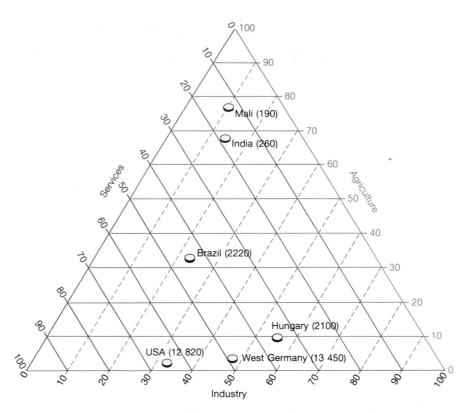

from country to country, and changes with time. The employment pattern by groups for six countries is shown in the triangular graph. Quite obviously the skills needed for employment in these countries differs a great deal, and the graph raises two questions. The first is whether there is any link between the proportions of people employed in the various groups and the average standard of living in the country. The second is whether the pattern of jobs available is due to the skills – or lack of skills – of people within the country, or to some other reason.

The employment structure (with per capita GNP in brackets) of six selected countries

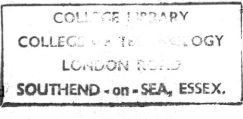
Some examples of the West African goldsmith's art

1 Which Third World countries were renowned in the past for a) very efficient crop growing using annual river flood-waters b) the invention of gunpowder c) providing the world's leading mathematicians and astronomers?
2 From the graph suggest a possible link between standards of living and patterns of employment – which patterns correspond with low standards?
3 Describe some wealth-producing jobs often done by foreigners in Third World countries.

Exercises on page 214
Zimbabwe

61

3.7 Unemployment in the Third World

Exercises on page 215
Comparative statistics for North and South.

In the industrialised countries of the North 'work' is usually thought of as employment for which salaries, wages or some other income is obtained. Many activities that can involve a great deal of time and energy, such as caring for a family or studying, are not paid for in the same way and are not regarded as real work or employment. Income from employment enables the worker and dependents to live in a certain manner. Standards of living vary a great deal because incomes from employment and other sources also vary a great deal. Employment can also give status and very often considerable job satisfaction. One of the big problems faced by countries of the North during the 1980s has been the growth of unemployment, with all the unhappy consequences.

There are many causes of increased unemployment in these countries. In part it is due to the changed nature of production, particularly greater automation and computerisation. Other reasons include increased competition for sales from other world producers together with a general decline in world trade. In some countries it is partly due to the decisions of their governments not to reduce unemployment if it means increasing the cost of living or having to

collect more taxes. Whatever the cause, unemployment can be demoralising for young and old alike. In the countries of the North, however, there is some financial support from the state and so most people can survive with a little security.

Unemployment in the Third World

In the Third World the scale and consequences of unemployment are far more severe. As shown by the diagram, a far bigger proportion of men and women who could work are either unemployed or underemployed. Although they could be trained to do a wide variety of jobs, there are far too few available in farming, mining, manufacturing and the service industries for the people that need them. One result of many people chasing few jobs is that wages are often low for the long hours worked, while working conditions and job security are very poor. Another is that jobs that would almost certainly be done by machines in the industrialised North are still frequently done by human labour in Third World countries. Such jobs are said to be 'labour-intensive'. While they may result in long hours of back-breaking toil, at least they provide employment and a modest income for millions of people.

The non-wage economy

In the Third World millions of people work hard to feed, clothe and gain shelter for themselves and their dependents without being in any paid employment. The extreme cases are subsistence farmers who literally live on what they produce, but many more manage to sell or exchange small quantities of their produce in local markets. Millions of children have to work long hours on family farms or in small workshops without a wage, while even more women have to combine their domestic tasks with unpaid farm labour.

Different from these unpaid workers

Basket-weaving in Rajasthan, India. Traditional craft industries still provide employment in Third World countries

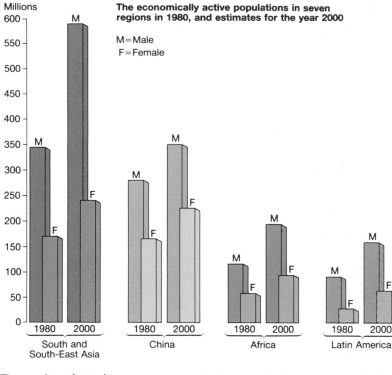

The economically active populations in seven regions in 1980, and estimates for the year 2000

M = Male
F = Female

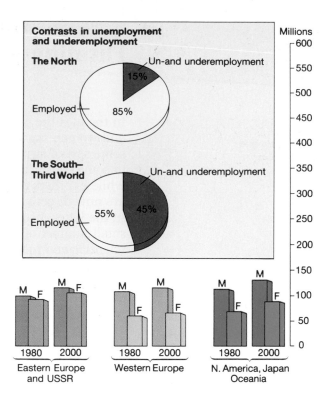

Contrasts in unemployment and underemployment

Millions

The North

Un- and underemployment

15%

Employed — 85%

The South— Third World

Un- and underemployment

Employed — 55%

45%

South and South-East Asia | China | Africa | Latin America | Eastern Europe and USSR | Western Europe | N. America, Japan Oceania

The number of people available for work, and unemployment in the North and South

This shoe-shine boy is part of the large informal economy in India

are the huge numbers, particularly in cities, who have irregular and casual work in the so-called 'informal' economy. These include people such as lottery ticket sellers, shoe-shine boys, water sellers, prostitutes, professional beggars and petty thieves. Their earnings are on or below the subsistence level, and are very uncertain. It is estimated that between 25 and 40 per cent of the labour force of many towns are forced to live in these highly insecure and marginal ways.

Future prospects for employment

Countries in the Third World have extremely high rates of unemployment and underemployment. From the diagram it can be seen that there will be an enormous increase in the numbers of people wanting jobs merely to survive by the turn of the century. These are not wild guesses, but accurate estimates made on the basis of children alive today. The vital question is whether more efficient and productive farming, manufacturing and tertiary activities will provide employment for these people. If not, are they doomed to an ever more precarious and declining standard of living?

Labour-intensive activity in China. These workers are building a reservoir

1 **a)** Put into words what the diagram predicts about changes in the number of people needing work by the year 2000. **b)** Why do the estimates for employable males and females differ so much?
2 Explain the changing nature of employment in your area, and if unemployment is high, describe some of the consequences for local people.
3 Why is unemployment a greater problem for people in countries of the South than of the North?

3.8 Farming

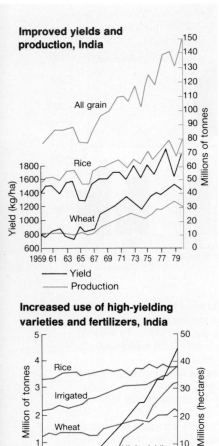

Improved yields and production, India

Increased use of high-yielding varieties and fertilizers, India

These two graphs show the effect of the 'Green Revolution' in India. Note the close relationship between the area planted with high-yielding varieties and the use of inorganic fertilizer

In the past the great majority of men and women spent their working lives in some sort of farming. We have seen that in some countries that remains the case, while in others only a small percentage are now directly involved in agriculture. In the latter, usually in the North, the small number of farmers use modern scientific and industrial methods to produce vast amounts of food. Indeed, one of the problems faced by these countries in the late 1980s is that they produce far too much for their needs, and have huge surpluses.

Farming in the Third World is very different. There is some commercial farming, with produce from large cattle ranches and plantations being sold within the country and exported, but most farming is small-scale and the produce enters only the local economy. This is not to say that farmers lack skills and understanding, but that they have not, or cannot, use the large-scale, mechanised and scientific techniques common in the North. With low living standards and a big increase in population quite certain, they have a desperate need to produce more food. At the same time there is an equally desperate need to reduce unemployment. For the sake of future generations it is also vital to achieve higher food production and a reduction in unemployment without damaging the agricultural environment that is the basis of survival.

The Green Revolution

Between 1940 and 1960 agricultural scientists working in Mexico developed new varieties of wheat and maize that gave far higher yields than traditional grains. Similar development of high-yielding varieties of rice took place in the Philippines. These new strains have been used with dramatic success in increasing production in Asian countries such as India, Pakistan, South Korea, Taiwan, the Philippines and Malaysia. In the rest of the Third World, apart from Mexico, Costa Rica, Colombia and Nigeria, the use of high-yielding grains has been more limited. The production of cereals in countries such as India, Pakistan and Mexico more than doubled over two decades due to the Green Revolution. It seemed that here was the answer to the problem of food production in the world. Since the middle of the 1970s, however, problems have arisen to be set against the increased production.

The high-yielding grains need large amounts of fertilizer to be fully productive. The global need for fertilizer is far more than can be provided, and most

Rice planting in Vietnam

A line of combine harvesters in Zambia

The Chinese example

Many a developing country could learn a lot from China. They could learn about, for instance, organic farming, with its emphasis on recycling of farm wastes for inexpensive fertilizer; about biological controls to keep down pests and weeds; about China's emphasis on human labour rather than costly machinery; about choosing crop varieties that suit local conditions; about soil conservation measures, notably terracing; about use of energy in forms other than fossil fuel, e.g. the methane gas produced by seven million bio-gas producers; about care for forests, which support agriculture through their watershed services and supply fuel (thus relieving pressure on valuable cattle dung); and about innovative aquaculture, enabling the Chinese to produce ten times more animal protein from one unit area of fish pond than the same unit of land given over to livestock.

Third World farmers are unable to afford it anyway. The growing crops are more susceptible to weed infestation, diseases and pest attacks than traditional varieties, and so more money is needed to apply herbicides, pesticides and fungicides. Traditional varieties do not provide such high yields, but can cope much better with extremes of drought or waterlogged soils. Unless a good irrigation system is available, crops of the high-yield varieties are more likely to be lost in extreme conditions. There is the additional fear that heavy applications of fertilizer and pesticides might lead to soil pollution and degradation.

Most of the benefits of the Green Revolution have been felt by richer, larger farmers who could afford the extra costs. Smaller farmers, tenant farmers, share-croppers and landless labourers have gained little. Indeed, they have often suffered as the better land has become concentrated in the hands of the more prosperous farmers. There has not been the expected increase in farm employment because production has become more mechanised. Critics of the Green Revolution accept the increased production, but argue that it has been at the expense of increased unemployment, concentration of land ownership, increased landlessness and a bigger gap between rich and poor.

The Chinese model

Adoption of farming methods found in the countries of the North is only one course that Third World farmers might follow. Before the Communist Revolution of 1949 millions of Chinese people starved each year, but now, with twice as many people, most are adequately fed. Much Chinese farmland looks like a garden landscape. Rice yields are higher than in India, mainly because of more irrigation and multiple cropping. The extract describes many of the features of the Chinese system. Some people express doubts about living under such a controlled system, but there is no question that the Chinese have found an alternative way to provide food and employment and also a caring attitude to their environment.

Exercises on page 216
World map showing share of agriculture in GNP

Fishpond harvesting in a Chinese village

1 With reference to the graphs, describe the increased yields and increased production of rice and wheat in India.
2 Draw up a table listing a) the benefits and b) the disadvantages of using high-yielding grains arising from the Green Revolution.
3 What do you think are the strengths and weaknesses of the Chinese model of agricultural production? Why might it not work in some countries?

3.9 Industrialisation

(Right) VW 'Beetles', assembled from German, Mexican and Brazilian parts, wait for distribution outside the Lagos assembly plant in Nigeria

The graph on page 61 showing the percentages of the workforce of certain countries in different types of employment suggests there is a link between manufacturing and GNP per capita. It seems that the more industrialised a country is the higher will be the average standard of living. If this is the case, then increased industrialisation should be the goal of countries in the Third World. Two important questions are whether the peoples of the Third World could provide the workforce for industry, and which types of industry would be of most benefit.

Types of manufacturing industry

There are many different types of manufacturing found in the Third World. Examples have already been given of long-established traditional craft industries (page 60). Long before contact with Europeans small workshops were producing textile, leather, wood and metal products of high quality for farming and domestic use. Contact with Europeans led to the import of cheaply manufactured goods from European factories,

A Mexican oil refinery

and the decline of much traditional craft. Since former colonies have gained their independence, however, many craft industries have revived and now play some part in the manufacturing industry of many countries.

Much industrial development consists of factories processing raw materials from farms and mines within the country. Examples include flour and sugar milling, meat packing and oil refining. Many of the products are for export and so the factories are often close to the coast and major ports, or have reasonable road or rail access to them. People in Third World countries have shown themselves quite capable of employment in these industries, and where there is a shortage of labour migrant workers are brought in from nearby countries.

There has also been a growth in some Third World countries of other sorts of manufacturing, not primarily based on local raw materials. These range from the manufacture of furniture and textiles, toys and games and printing and publishing to huge steel works, car factories and shipyards. Sometimes the capital for these industries is provided by wealthy local businessmen, but many are part of huge multinational companies. Others are owned by the state, and are developed as part of centrally-controlled government planning for the country.

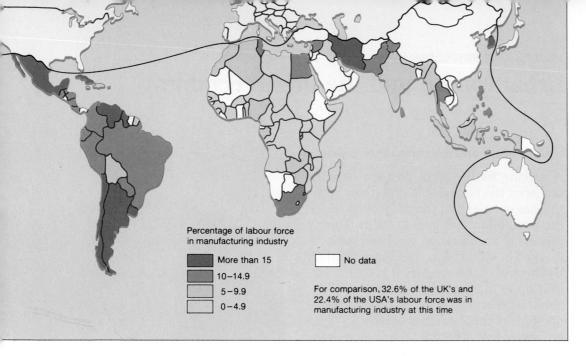

Percentage of labour force in manufacturing industry

- More than 15
- 10–14.9
- 5–9.9
- 0–4.9
- No data

For comparison, 32.6% of the UK's and 22.4% of the USA's labour force was in manufacturing industry at this time

The percentage of the labour force in manufacturing industry in Third World countries

Contrasts in industrialisation

From the map it can be seen that there are big differences in the percentage of the labour force in manufacturing in the Third World. Highest proportions are in countries such as Brazil, Mexico and Argentina in Latin America, and Taiwan, South Korea and the Philippines in the Far East. These are sometimes called the Newly Industrialised Countries, and they are now beginning to compete not only in the Third World but also in countries of the North. Steel and ships, electronic goods, cars and consumer goods are manufactured in these countries using local employees, and sold around the world. Many multinational companies have located factories in such countries because of the efficient but relatively cheap and non-unionised labour force.

By comparison many countries, particularly in Africa, have little industry. Often there has been some very expensive massive new project, such as a hydro-electric power station or a steelworks, that contributes to the economy but provides little employment or income for the majority of the population. In most Third World countries the smaller-scale workshop production goes on side by side with, but independent of, the huge state or multinational industries.

People and industrialisation

Modern industries need money, whether from wealthy local businessmen, state taxes or foreign sources. There is also need for energy supplies and good road, rail, air and water transport. But there is also need for an efficient workforce. It isn't easy to change from an agricultural way of life to one in factory or mine. It involves a lot of training and adjustment. All the evidence from the Newly Industrialised Countries is that people in Third World countries are perfectly capable of working in modern industry, not only as unskilled labourers but as expert technicians and skilled managers. If the employees are well paid, they then in turn increase the demand for goods in the home market, leading to further growth. But with manufacturing as with farming, increased production must be matched by increased employment, and resource waste and environmental damage must be avoided. Large-scale 'Western' type of industry may not be the appropriate technology for all Third World countries.

1 With reference to the map, name one country with more than 10 per cent and one country with less than 10 per cent of its labour force in manufacturing industry in Latin America, Africa and Asia. Which continent is least industrialised?
2 Why might industrialisation not lead to a) more employment b) better standards of living?

For world map showing levels of industrialisation, see page 114

Exercises on page 217
Zambia

3.10 Urbanisation and the urban economy

Urbanisation and migration into towns

A striking feature of the past few decades has been the remarkable growth in both the size of cities and the proportion of the population that lives in them. This process of urbanisation has been most marked in countries in the Third World. By the end of this century it is estimated that up to three-quarters of the world's population will be living in urban areas, one-fifth of these in cities of over 1 million people. The five largest cities will have populations of over 20 million, with the largest, Mexico City, nearer 30 million. Of the 60 or so cities with over 5 million people, 45 will be in Third World countries.

A street scene in Old Delhi, India

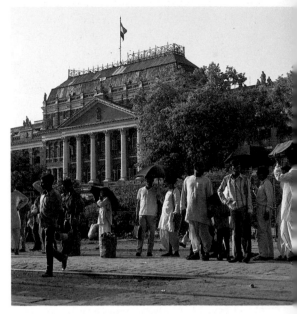

Commuters in Calcutta, India. Commuting to and from work is a part of life for millions of people in cities all over the world

The 'dual economy' of Third World cities

The great differences in housing and living standards are also reflected in the jobs people do. Different labels are used for the two distinct types of economy that have evolved. Sometimes they are called 'formal' and 'informal', sometimes 'higher circuit' and 'lower circuit'. Whatever the labels, the differences are very real.

The upper circuit economy consists of large-scale companies which are often foreign-owned. Manufacturing usually uses the modern techniques used in the developed countries and products tend to be for the overseas market or the wealthier members of the population. Also in this group are the business and commercial activities, and the jobs provided by the government. It is calculated that between 10 and 15 per cent of the employed population of Third World cities work in these types of jobs, and do them very well. Because there is such a shortage, however, those that do have them usually work long hours, are often exposed to dust, excessive noise and dangerous chemicals and machinery, and get low wages.

The lower circuit economy consists of thousands of small locally-owned factories and workshops. Although the firms are small, their vast number means they employ a lot of people in total. Prices and profits are low, but employment for members of the family may matter as much as profits alone. Apart from these small workshops and bazaar activities that provide a precarious livelihood for huge numbers of poorer people, there are those activities which provide casual and semi-employment as described on page 63.

One reason for this growth of Third World cities is natural increase, with births exceeding deaths. Another very significant reason is the massive in-migration of people from the rural areas and smaller towns. Although wages are on average very low, and unemployment high, chances of some income and means of survival are better than outside the cities. So every day tens of thousands of people move into the cities seeking work and a share of some of the services found in urban areas. Sadly, many are unsuccessful. They find themselves in large shanty towns on the outskirts of the cities, living in squatter houses of corrugated iron, packing cases and plastic sheets. Clean drinking water and sanitation is often unavailable, food and fuel is scarce, and the vast majority must do without electricity and medical care.

As we saw on pages 10 and 16, conditions in the squatter settlements fringing most Third World cities can vary a great deal. In a few, some amenities are provided by the local city authorities, and some families will slowly build more substantial homes for themselves. Nevertheless there remains a big gulf between the relatively wealthy who live in housing areas as good as anywhere in the North, and the great majority of the population who live in conditions ranging from modest but bearable to degrading. Contrasts in urban living standards are greater than in the cities of either the developed 'market economies' or the East European and Soviet socialist countries. It is as though there are two different ways of life within the one urban area.

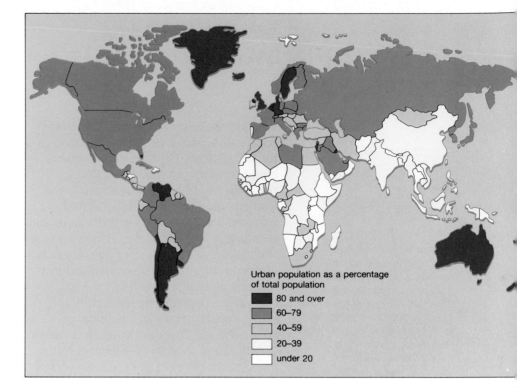

Urban population as a percentage of total population

- 80 and over
- 60–79
- 40–59
- 20–39
- under 20

The percentage of people living in towns and cities, by country

Urbanisation and development

Increased urbanisation might be considered a sign of increased wealth and improvement in living standards. But there is no such simple link. Urbanisation in the Third World is not necessarily linked to industry, business and the creation of wealth, nor does it lead to standards of living that can be properly described as developed. But it is not that people in Third World countries are unable to do urban jobs. The explanation lies rather in the nature of Third World urbanisation.

1 Give examples of three sorts of weatlh-creating jobs outside cities, and three normally found within cities.
2 Explain the world pattern of urbanisation. What does the map *not* tell you about world urbanisation?
3 Explain why the relationship between the wealth of a country and the process of urbanisation is not a simple one.

Exercises on page 218
The world's largest cities

A bazaar salesman in Delhi

69

3.11 Women

Women make up about half the world's population, and a similar proportion in most countries. On that basis alone it could be expected that work and its rewards would be equally shared, and that the quality of life of males and females in any particular country would be more or less the same. This is far from the case, and in most countries there is discrimination against women in terms of work available, rewards for work, ownership of property, legal rights, social opportunities and political responsibilities. This discrimination varies greatly from country to country, but a few differences exist in countries of both the North and South.

The work of women and their rewards

Women all over the world have traditionally not only raised their very young children, but also sustained their families by providing food and care and looking after their health. In some societies this was possible as a genuinely shared partnership, but all too often in the past women were subservient to and dependent on men in the family. In many societies today, especially where they find it hard to be economically independent, women seem little more than unpaid servants. They work very hard to support the family of children and a husband who may be a wage-earner or working for himself. If they are fortunate they will receive the support and help of their partner and even a share of the income. All too often they may be abandoned to cope with the family as best they can.

In addition to their domestic work many women have to work on the land or in factories to add to the family income. Women often work long hours on the land but for reasons of tradition have no right to any income from the sale of produce nor to ownership of the land. It is estimated that in Africa as a whole women do 60 per cent of the agricultural work, 50 per cent of the caring of animals and all the food processing. In addition they are usually responsible for collecting firewood and water. Women make a great contribution to the development of countries through this unpaid labour.

When it comes to wage-labour, women in the Third World suffer the same disadvantages as women everywhere. Some types of work are not available to women, despite evidence that most jobs could be done by women if given sufficient training. When jobs are available to both men and women, women frequently receive smaller wages for the same task. It is also usually much more difficult for women to attain positions of authority and leadership when both sexes are employed.

Women do much of the labouring in rural Africa. (Below) what does this diagram *not* tell you about the work that African women have to do? (Bottom) a typical day for a Zambian woman, who is likely to be undernourished during the busiest times of the year because she is simply too busy to feed herself properly

Women's work in Africa

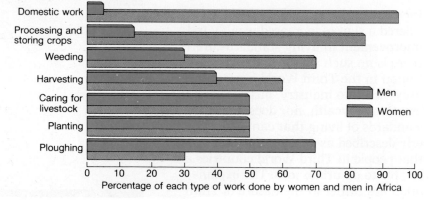

Percentage of each type of work done by women and men in Africa

The working day of a Zambian woman

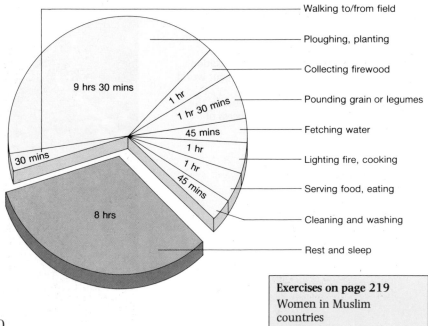

- Walking to/from field
- Ploughing, planting
- Collecting firewood
- Pounding grain or legumes
- Fetching water
- Lighting fire, cooking
- Serving food, eating
- Cleaning and washing
- Rest and sleep

Exercises on page 219
Women in Muslim countries

Equal pay ? No, women get paid less than men

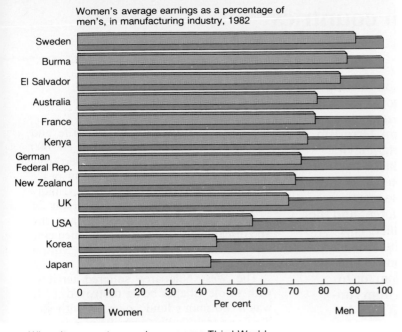

Women's average earnings as a percentage of men's, in manufacturing industry, 1982

Sweden
Burma
El Salvador
Australia
France
Kenya
German Federal Rep.
New Zealand
UK
USA
Korea
Japan

0 10 20 30 40 50 60 70 80 90 100
Per cent

☐ Women   Men ☐

When it comes to equal pay, some Third World countries do better than major developed countries

Female workers in Colombo, Sri Lanka

The rights of women

The legal status of women varies around the world. This shows in such things as whether they are allowed to own property, borrow money, vote in elections and so on. In some countries women are still not allowed to vote, and in none is there an equal share of political representation and power. A few women have held the highest political office in countries such as Sri Lanka, India, Israel, the United Kingdom and the Philippines but, world-wide, women are underrepresented in government at all levels. Women are slowly gaining their political rights, but they still remain a long way off for many.

Both cause and consequence of discrimination against women is difference in educational opportunity and attainment. There are remarkable variations in school enrolments and adult literacy around the world that can only be explained by cultural traditions and traditional practice. Many see equality of educational opportunities as a vital step towards real equality between men and women.

It is obvious that in many societies an enormous amount of talent and ability is not being used because of discrimination against women. Fur-

Education in six countries compared

Country	GNP per capita (US$) 1985	Primary school enrolment ratio (per cent) 1981		Secondary school enrolment ratio (per cent) 1981		Adult literacy rate (per cent) 1985	
		male	female	male	female	male	female
Malawi	170	70	49	6	2	52	31
Pakistan	380	73	30	22	8	40	19
Brazil	1 640	93	93	29	35	79	76
Cuba	1 526*	100	100	73	80	96	96
Australia	10 830	100	100	85	88	100	100
United Kingdom	8 460	100	100	81	84	99	99

*Net Material Product

thermore, women often work extremely hard for unjust rewards and an inferior quality of life, whatever the general level. Economic development needs the talents and contribution of women, while social development means that the rewards for their efforts should be equal to those for men.

1 Describe what you find surprising and not surprising about the data for work done by women in Africa.
2 Calculate the time spent on domestic work, other work (mostly non-paid), and time for leisure and sleeping for the Zambian woman. Compare this with the day of one or more women with families that you know.
3 Describe and try and account for some of the educational discrimination against females shown in the table.

Making 'jera', or Ethiopian bread

3.12 Conflicts within countries

One widespread constraint on development is conflict between people within a country. Such disharmony may show as violent protest, harsh control, terrorism or guerrilla fighting or outright civil war. People are wounded, killed or imprisoned, services disrupted and property destroyed. Valuable resources are spent by police, armed forces and the groups involved in the conflict. This violence may well be seen by the different sides as necessary to control terrorism, to gain justice or freedom, or to prevent an opposing group from running the country. The result may be greater justice for the majority of the people, and so a more developed society, but during the conflict there is bound to be a great drain on material and economic resources.

When it comes to armed conflict, events change rapidly and this map may no longer be accurate when you come to read it

Ugandan terror

Uganda has suffered a lot of strife and bloodshed in recent years. About 800 000 people fled in 1979 following the overthrow of Idi Amin. As some semblance of peace appeared, people returned in droves, many of them in poor health as a result of arduous camp life. Malnutrition was rife and disease flourished. Food and medicine were in short supply. Then came Milton Obote's reign of terror ...

The West Nile area suffered terribly, with whole villages being destroyed ... It was a 'no man's land', overrun with soldiers and troubled by guerrilla movements ... People lived in continual fear.

Security remained precarious until Museveni came to power in January 1986, impressing both Ugandans and the outside world with his disciplined and courteous soldiers, many no more than children. A relative calm exists, and the main Kampala–Arua road is open once again. With the declared commitment of Museveni to bring about peace a semblance of normal living has returned to Uganda.

Extremes of wealth and poverty

Conflicts may arise when one section of society feels it does not get fair rewards for its efforts nor a say in how it can live. This is likely to happen where there are great contrasts in wealth and living standards within the country, and where the wealthy own most of the land and industry and have great control or influence over the government. In a democracy, governments carrying out unpopular policies or supporting injustices can be voted out of office if enough people so desire. In dictatorships, or democracies where elections are controlled by those already in power, such peaceful change is impossible. In these cases violence becomes the only way of opposing injustice, exploitation or brutal control. There are also bitter internal conflicts between groups who have strongly different political views on on what their society should be like.

Wars in Africa

A quarter of the world's nations are currently caught up in conflicts which have already claimed up to 5 million lives. More than 4 million troops are fighting each other. Most of these wars are in the Third World, many in Africa. Many are encouraged or supported by superpowers or other nations. Religious, tribal and political differences are the main causes.

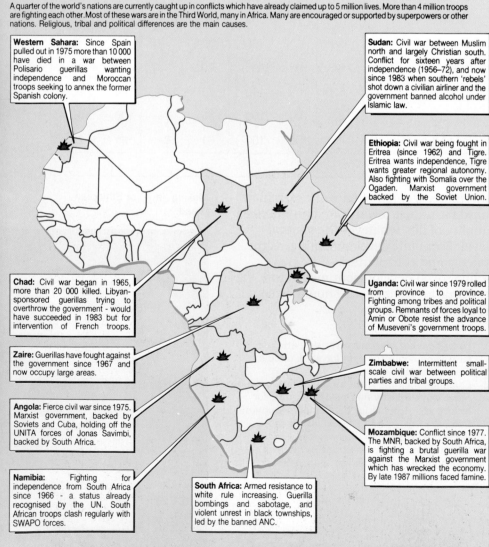

Western Sahara: Since Spain pulled out in 1975 more than 10 000 have died in a war between Polisario guerillas wanting independence and Moroccan troops seeking to annex the former Spanish colony.

Sudan: Civil war between Muslim north and largely Christian south. Conflict for sixteen years after independence (1956–72), and now since 1983 when southern 'rebels' shot down a civilian airliner and the government banned alcohol under Islamic law.

Ethiopia: Civil war being fought in Eritrea (since 1962) and Tigre. Eritrea wants independence, Tigre wants greater regional autonomy. Also fighting with Somalia over the Ogaden. Marxist government backed by the Soviet Union.

Chad: Civil war began in 1965, more than 20 000 killed. Libyan-sponsored guerillas trying to overthrow the government - would have succeeded in 1983 but for intervention of French troops.

Uganda: Civil war since 1979 rolled from province to province. Fighting among tribes and political groups. Remnants of forces loyal to Amin or Obote resist the advance of Museveni's government troops.

Zaire: Guerillas have fought against the government since 1967 and now occupy large areas.

Zimbabwe: Intermittent small-scale civil war between political parties and tribal groups.

Angola: Fierce civil war since 1975. Marxist government, backed by Soviets and Cuba, holding off the UNITA forces of Jonas Savimbi, backed by South Africa.

Mozambique: Conflict since 1977. The MNR, backed by South Africa, is fighting a brutal guerilla war against the Marxist government which has wrecked the economy. By late 1987 millions faced famine.

Namibia: Fighting for independence from South Africa since 1966 - a status already recognised by the UN. South African troops clash regularly with SWAPO forces.

South Africa: Armed resistance to white rule increasing. Guerilla bombings and sabotage, and violent unrest in black townships, led by the banned ANC.

In the 1980s many conflicts have been due to these causes. In Peru, for example, there is a violent struggle between communist revolutionaries and the democratically elected government forces – with many people in the Andean villages and the cities suffering at the hands of both. In Nicaragua a revolt of oppressed people led to the overthrow of a harsh dictatorship, but civil war continues between the government and its supporters and counter-revolutionary supporters. The Philippines saw a remarkably peaceful overthrow of a President who seemed to hold on to power in a democracy by corrupt means. The new government has to cope, however, with a potential conflict with communist guerrillas.

Religious rights and self-determination

Some religious and ethnic groups feel they are unjustly treated, or are actually persecuted, purely because their religious beliefs, race or cultural backgrounds are different from those of the people holding power. Some conflicts within countries are due to these differences, or some combination of them.

In the past powerful nations frequently took land from less powerful groups, sometimes by force but often by dubious treaties. Today in countries such as the United States of America, Australia and New Zealand descendants of original Indian, Aboriginal and Maori groups are attempting to reclaim some of their rights to lands that were taken by the invaders. They are too few and weak to engage in violent protest, but increasingly have moral and practical support from others in their countries who recognise they have been badly treated.

Elsewhere some of the most violent conflicts relate to religious, cultural and tribal groups demanding rights to what they claim is their territory. Basques in Northern Spain, Tamils in Sri Lanka, Catholic Irish in Northern Ireland, Sikhs

These Ugandan 'boy soldiers' have been swept up in an internal conflict that has lasted nearly twenty years

in the Punjab in India and Palestinians in Israeli-occupied territory are but a few of the groups seeking their rights to self-determination. In most cases there are those who feel so strongly about their situation or views that they have resorted to violence to get their way. They are then challenged by the government with equal violence and so the conflict grows until one side is defeated or some compromise is reached. In all these cases the rights and wrongs of the cause and action are passionately argued. Whether the conflict leads to greater justice and development remains to be seen.

1   Read the extract and the description on the map about Uganda. What were the causes of conflict? What were some of the damaging effects of the conflict on the country?
2   The map highlights various conflicts in Africa. For each conflict, using the description or other evidence you have, say whether they are mainly religious, tribal, racial or political. What evidence is there of outside interference and involvement?
3   Do you think violence by a) governments seeking to control opponents and b) extremist groups claiming to represent oppressed groups within a country is ever justified?

Exercises on page 220
Conflicts around the world

Aboriginal demands for traditional land rights have gathered pace in recent years

3.13 South Africa and apartheid

One of the most notorious examples of conflict within a country based on racial, cultural and economic differences can be seen in the Republic of South Africa. The lands of South Africa have been settled over the centuries by different African tribal groups such as the Swazi, Zulu and Xhosa as well as the original Bushmen and Hottentots. Permanent settlements were also established by the Dutch, later to be known as the Boers, and the English. Many Asians, especially from India and Malaya, were employed on White-owned farms and plantations. There were frequent wars between the White and Black settlers, as well as between the English and Boers, before the present pattern of population distribution and type of economy was established.

Africans make up about 74 per cent of the total population, with Asians and Coloured groups contributing a further 11 per cent. The combined English- and Afrikaans-speaking White group makes up about 15 per cent of the total. Some of the features of the different racial groups are given in the table. Everyone has to be officially classified into one of these groups, and that determines where you can live, the education and health services you receive, the type of work you can do and the range of salary received, voting rights (if any), and until recently who you could and could not marry. The decision to classify people in this way, and to pass laws compelling groups to live in different ways, was made by the Whites. Although small in number the Whites are by far the most wealthy, privileged and powerful group in South Africa, with absolute control of the police, armed forces and means of enforcing the law.

'Colour coding' – racial groups in South Africa

Colour coding

Racial category		Population	Per cent of total
African	'Pure' African, member of one of nine tribal or 'national' units–North Sotho, South Sotho, Tswana, Zulu, Swazi, Xhosa (two units), Tsonga and Venda.	24 103 458	73.8
Coloured	'Mixed race'. This means not only the descendants of black-white liaisons centuries ago and more recently but also acts as a catch-all for anyone with light brown skin not designated 'Indian'. Some 'coloured' people look 'white', others look 'black'.	2 830 301	8.7
Indian	Usually refers to people whose ancestors were brought from the Indian subcontinent by the British to work on railway construction etc.	890 292	2.7
White	People whose ancestors were of European origin, most of whom speak either Afrikaans (an offshoot of Dutch) or English. Most white South Africans probably have some 'mixed blood' at an average recently estimated to be seven per cent.	4 818 679	14.8

South Africa, with its black 'homelands'

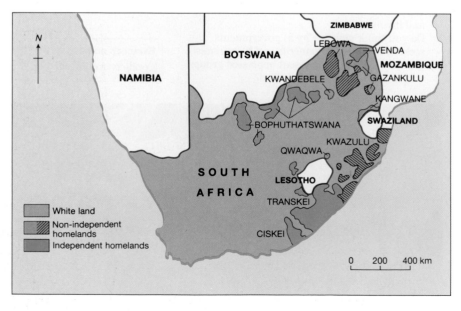

Homelands and townships

The Land Act of 1913, passed by the newly-allied Boers and British, removed from Black people the right to own land. It gave 87 per cent of the land to Whites. The rest, which was mostly fragmented and scattered parcels of some of the least fertile farmland, lacking in mineral and water resources, was given as 'reserves' to Black people. In 1948 the government decided to introduce a policy of apartheid, or 'separate development'. The basic idea was to create ten homelands from the reserves, and to make every African a citizen of one of them on the basis of their tribal links. The result has been to allocate millions of Africans to tribal 'homes' they have never visited, and that would be unable to support them at a reasonable standard of living should they all try to live there. The homelands are intended to attain independence, and four were considered to be independent by the South African government and their own leaders by 1986 – but millions of Africans in South Africa and all other countries around the world refused to recognise them as countries.

One complication is that large numbers of Africans are needed in the White areas to work on farms, in mines and factories and in various services. These Blacks have to live in special areas, usually large townships, that have been built some distance from the White urban areas. Many hundreds of thousands of mineworkers are temporary migrant labourers from neighbouring states such as Lesotho, Swaziland, Botswana and Zimbabwe. These migrant workers live in special mining compounds, segregated from the townships and the white areas.

Inequality and oppression

In terms of resources, South Africa is one of the richest countries in the world. Although the Black people of South Africa may be as well-off as those in some other African countries they do experience very great poverty in relation to the Whites. Many Blacks resent their lack of social and political rights and refuse to accept the idea of apartheid. They want a fairer share of South Africa's wealth. Since their leaders are in prison and they have no way of peacefully expressing their views, many have turned to violence and rebellion. In turn the government has used force to control the unrest, and in 1986, 1987 and 1988 ruled the country under a state of emergency. There was growing violence between Blacks and Whites and between different Black groups. So threatening had the situation become, countries from all over the world were debating the possible use of sanctions to force the government to change from its policy of apartheid. Because of it millions suffer poverty and oppression, while the conflict and its control means the spending of vast sums of money and the use of valuable resources.

1. What signs of poverty and differences in living standards are mentioned in the advertisement? What is the link with apartheid? How reliable are the organisations that produced the advert?
2. Which homelands had been declared independent by 1986? Why do so many Black Africans reject apartheid and the idea of their own nation states?
3. What is the main principle of apartheid? Is it right for outside nations to interfere with South African policy?

Attempts to move squatters from this township near Cape Town have resulted in violence

Exercises on page 221
Population statistics

An advertisement by Christian Aid and Oxfam

THE LINKS BETWEEN
APARTHEID AND POVERTY

Christian Aid and Oxfam have worked for many years in South Africa. In our experience apartheid is a major cause of poverty

Poverty is institutionalized by South Africa's apartheid system. It underlies a mortality rate five times greater among black children than among white. In a land rich in natural resources, some 10% of children born in black areas die of malnutrition.

Although black people comprise about 85% of the population, they are denied basic rights, and thus have no means for achieving a just share of educational resources or medical services. 3.2 million black people have been forcibly removed since 1960, under the apartheid system, to the barren soil and overcrowded settlements of so-called 'homelands.'

Our South African partners want the whole of the international community to put effective pressure on the South African Authorities to end apartheid.

Fact and figures behind apartheid and poverty:	Black	White
Population of South Africa (%)	85	15
Distribution of land (%)	13	87
Average monthly earnings (Rand)	320	1 350
Education: expenditure per child/year (Rand)	238	1 654
Health: population per doctor	40 000	400

3.14 Refugees from conflict

(Right) these Palestinians in Beirut belong to one of the largest refugee groups in the world

People are sometimes forced to leave their homes because of the devastating effects of natural disasters such as earthquakes, floods or droughts. Very different from these are the millions of people who through fear of persecution and ill-treatment due to their race, religion or nationality, or to avoid being caught up in civil war, flee their country. In the mid-1980s there were about 10 million refugees of many different nationalities scattered all over the world. Their standards of living vary enormously, but none can manage unless the governments and people of other countries are willing to give them help and protection.

There have been political refugees since the beginnings of known time. In earlier centuries persecuted people could move to another part of the world and establish a new homeland, but as the world population grew and societies became more organised it became more difficult for refugees to settle in a foreign country. The United Kingdom is one of a number of countries with a long reputation of helping political refugees from all over the world. A general idea of the numbers and distribution of refugees is given by the world map.

The more prosperous countries are able to support these newcomers without too much strain on resources or too much social tension. Indeed they often make considerable contributions to the life of the host country through their skills, hard work and rich cultural traditions. But there are parts of the world where large refugee populations add to already very great difficulties. In 1950 the United Nations set up the Office of the United Nations High Commissioner for Refugees, a non-political organisation committed to protecting and helping refugees in every corner of the world. There are other UN agencies to help the two million Palestinian refugees and the quarter of a million Kampucheans in the border area of Thailand.

The crisis areas

Some of the more difficult refugee areas are shown on the map. The two million Palestinians were forced to leave their homes in 1948 as a result of the creation of Israel and the ensuing conflict between the new state and neighbouring Arab countries. Many of the original refugees have died, and now the majority were born in a foreign state. Most live in difficult economic circumstances, and remain bitter about the loss of what they regard as their homeland. Continuing guerrilla or open warfare between the Palestinians and supporting Arab

Amerindian refugees from Guatemala in a Mexican camp

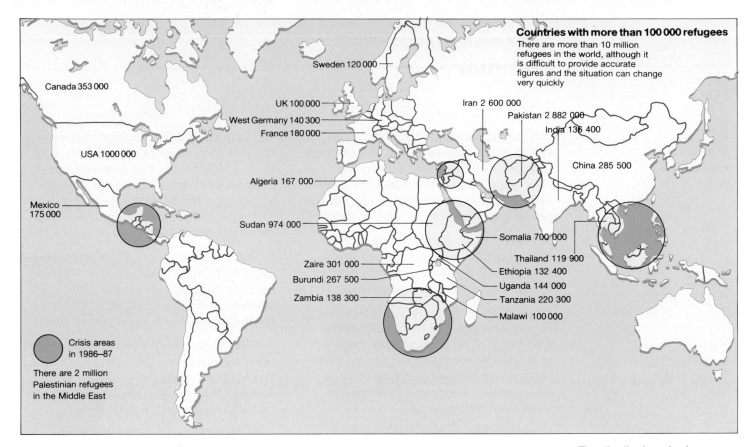

Sweden 120 000

Canada 353 000

UK 100 000

West Germany 140 300

France 180 000

USA 1 000 000

Iran 2 600 000

Pakistan 2 882 000

India 136 400

China 285 500

Algeria 167 000

Mexico 175 000

Sudan 974 000

Somalia 700 000

Thailand 119 900

Zaire 301 000

Ethiopia 132 400

Burundi 267 500

Uganda 144 000

Zambia 138 300

Tanzania 220 300

Malawi 100 000

Crisis areas in 1986–87

There are 2 million Palestinian refugees in the Middle East

The distribution of refugees around the world

countries and Israel add to all the other problems faced by these refugees.

There are over four million refugees from Afghanistan, where until 1988 a large military force from the USSR was fighting alongside the government against rebels opposed to the government. The refugees from this civil war live mainly in camps in Pakistan, Iran and Turkey, although some have now moved away from the camps to earn a living elsewhere in these countries.

In South-East Asia large numbers of refugees continue to flee from Kampuchea and Vietnam, including the 'Boat People' who risk shipwreck and pirate attack to reach safety. Other equally tragic refugees are found in Central America. To escape persecution and fighting of the most savage kind between supporters of the governments and rebel groups in El Salvador, Nicaragua and Guatemala, they seek asylum in Mexico, Honduras and the USA. The plight of some is to be refused entry to these neighbouring countries and to be sent back to face fresh horrors.

Almost half the world's refugees are in Southern and East Africa. Poor countries here find it hard to cope with the arrival of large numbers of refugees when many of their own people are

struggling to survive. Many African countries are nevertheless generous to refugees, and help by providing emergency camps and encouragement to settle and earn a living. In some cases, such as Ethiopia and Sudan, there is a complicated and continual flow of refugees across the border in both directions.

Refugees usually need immediate support in terms of food, shelter, clothing and medical care. But they also need protection and the security of knowing they will not be sent back to the country they have left, nor be separated from their families. In the longer term they need support for education, health care, employment and a chance to be responsible for their own lives. Refugees are a symptom of lack of development in their own land, and an additional problem in the economic and social development of most host countries.

1 **a)** Name two more prosperous and two Third World countries that have large numbers of refugees. **b)** What are the differences and similarities in the help that can be given to refugees by prosperous and less prosperous countries?

2 Describe a group or groups of refugees in Britian who have come because of persecution due to race, religion or nationality.

Exercises on page 222
Ratio of refugees to local population: top twenty countries

3.15 Summary

☆ There is no simple relationship between population numbers, density and standard of living within a country. Over-population exists when the available resources are unable to support the people regardless of the total or density.

☆ Where living standards are already low, more children can further strain resources. But children may be seen as future sources of help and security.

☆ Health care in many Third World countries is poor. Ill-health affects the well-being of people and their capacity to work. Improvements in provision of clean water, proper diet, hygiene and basic medical care are a vital requirement for development.

☆ In the Third World there is a huge need for better and more widespread education: basic literacy for everyone, and training in the skills required for economic and social development. Education is needed for development, but resources are lacking.

☆ There is need for a balance between the provision of large capital-intensive projects (like dams and airports), and support for more widespread, smaller-scale 'appropriate' technology and development in farming and manufacturing.

☆ There is much un- and underemployment in Third World countries. The result is migrant labour, dispersed families, and the exploitation of cheap labour, especially women and children.

☆ Women are discriminated against in most societies, and this is so in the Third World. They are often not allowed to develop and use their skills, and are unfairly rewarded for their work.

☆ In Third World countries, as elsewhere, many people are undervalued and exploited by those with power and authority, and wealth is not fairly shared. This can lead to resentment and conflict.

☆ Many people are discriminated against or persecuted because their religion, race or nationality is different from the majority or from those in power. Resulting conflicts are both a symptom and a problem of underdevelopment.

☆ Large numbers of refugees originate from and are supported by other Third World countries. They are fleeing natural disaster or persecution.

☆ People in Third World countries have the potential to develop the skills necessary for development. Lack of development cannot be blamed on lack of potential ability of any people.

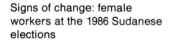

Signs of change: female workers at the 1986 Sudanese elections

Chapter 4

Development and dependency

'Food for Work' aid, Mali

Third World countries are part of global economic and political systems. To what extent is their relative lack of development due to past exploitation by other countries, or present external political, economic or financial control or influence?

4.1 Colonies and Empires

colony *n*. settlement or settlers in new country forming community fully or partly subject to mother State; their territory; people of one nationality or occupation in a city, esp. if living more or less in isolation or a special quarter ...

empire *n*. supreme and extensive (political) dominion; absolute control (*over*); government in which sovereign is called emperor; territory of an emperor; ...

The Concise Oxford Dictionary

A contemporary illustration showing the ill-treatment of the American Indians by the Spanish

It is often claimed that poverty and lack of development in most Third World countries is less to do with harsh environment, few resources and inadequate people than with the fact that they have been occupied and exploited by other nations. Although many countries that were colonies and parts of empires are now independent states, they are still suffering from their past experiences and the consequences still affect them today.

No one knows for certain when and where human groups first appeared on the earth, but evidence from archaeology, visible evidence and written records suggest that several thousand years ago there were many distinct cultural groups scattered around the world. Evidence of early 'civilisations' has been found in the Andes of South America, in Mexico, along the Nile river in Egypt and the Tigris-Euphrates rivers in what is now Iraq and Iran, in West and East Africa, in India, South-East Asia and in China, and in a number of places in Europe. For a variety of only partially understood reasons, a number of countries in Europe became particularly powerful some five or six hundred years ago, and began to explore and expand into the rest of the world. They did this for a number of reasons, including curiosity, a wish to spread the Christian religion, and to gain wealth and territory through trade or possession.

With a mixture of skill, courage and endurance, frequently with disdain or contempt for people of different race or culture, and with superior fighting power, the Spanish, Portuguese, Dutch, French, British, Italians and Germans 'discovered' new lands, settled in them, traded, and took possession of them by treaty or force. By the end of the last century much of the world was controlled by European countries either directly as part of their empires, or indirectly through the power held by colonial settlers from the home nation.

Spaniards in South America

The story of Christopher Columbus sailing westwards across the uncharted Atlantic until reaching the Caribbean islands, at that time unknown to Europe, is well known. When news of these lands and their wealth was brought back to Europe, many more, particularly Portuguese and Spanish, set out to seek their fortune, claim land for their country, and to spread the Christian faith. To prevent quarrels between Spain and Portugal the Pope decreed that all land to the east of a line drawn from North to South Pole, and running through South America, should be Portuguese and the rest Spanish!

Spaniards moved into Mexico and began exploring what is now the west coast of South America. A remarkable group of people known as the Incas controlled a huge mass of land in the Andes stretching from what is now Ecuador to northern Chile – about the same distance from southern Britain to

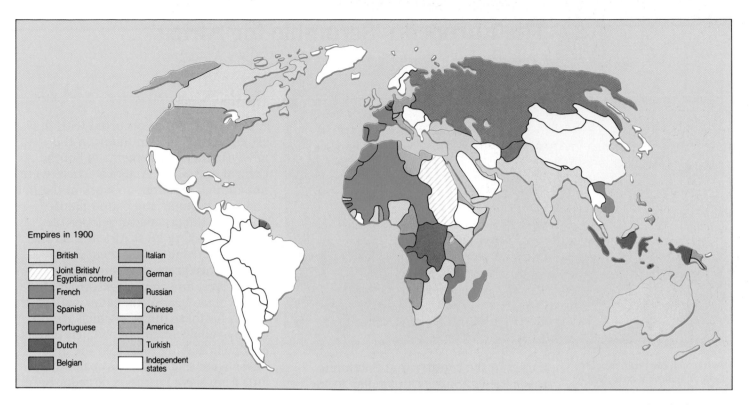

The main empires in the world, 1900

Empires in 1900

- British
- Joint British/ Egyptian control
- French
- Spanish
- Portuguese
- Dutch
- Belgian
- Italian
- German
- Russian
- Chinese
- America
- Turkish
- Independent states

the southern Sahara desert in Mali. From records kept by the early Spanish we know that Inca people were born to a place in society for life, as farmers, craftsmen, architects, soldiers, priests, commanders and so on. Although some of their customs appear savage to us, provided they worked and fought hard for the Lord Inca they were materially quite well-off, and were cared for by the state when ill and elderly. The Inca people were skilled craftsmen, excellent architects and masons and engineers. They had mastered the art of governing an empire and used the resources of their mountain environment very cleverly.

Early in January 1530 the Spaniard Francisco Pizarro landed on the coast with 180 men, known as conquistadors, and 20 horses. Armed with steel swords, armour, crossbows, muskets and cannon they moved inland. Over the next two years, against great odds, they conquered many armies, took more and more land, and finally vanquished the Inca leader and his army, putting him to death for refusing to become a Christian. Huge sums of gold collected as ransom were sent back to Spain, and a period of Spanish domination and exploitation of the land and its people began. Some idea of the consequences of the Spanish invasion and conquest are given in the extract.

This is one of the earliest examples of a European nation creating an Empire in a different continent. Many other examples were to follow in the succeeding centuries. As the map shows, most Spanish and Portuguese colonies had gained their independence by the year 1900. But much of Africa and Asia was still part of European empires, and it was many more decades before they became independent states. There is no doubt that the experience of foreign domination affected all these countries, sometimes for good, but often for bad.

1 Name one country in South America that was controlled by Portugal, and one by Spain. How will this be evident in these two countries today?
2 Argue the cases for and against the rights of a country to have an empire.

> **Exercises on page 223**
> South America

An extract from *Pizarro and the Conquest of Peru* by Cecil Howard

In Peru

For the first time Charles V (of Spain) held undisputed sway over his new realm in South America. Riches poured into his treasury and he appointed governors to enforce his laws. But although the authority of the Spanish Crown was not seriously challenged for almost three hundred years, royal attempts to control the unruly New World colonists never were completely successful. And the natives of Peru themselves, a gifted, industrious, and law-abiding people, who under a more enlightened rule than that of the conquerors might have led the march of civilisation in South America, instead sank into a poverty and despair from which they have scarcely emerged to this day.

4.2 The European 'Scramble for Africa'

Ashanti chiefs (from modern Ghana) submit themselves to their British conquerors, 1895

The slave trade

A profound change occurred with the establishment of plantations in the newly-founded settlements of South, Central and North America. There were not enough labourers to work in the hot and humid sugar and cotton plantations, so the practice developed of forcefully taking West Africans across the Atlantic and selling them as slaves. The practice grew up of shipping goods out to West and south-west Africa that the local kings and chieftains wanted. In return for these the local chiefs would make available men and women taken captive in slaving raids into the interior. These were sold to the European merchants from slave fortresses dotted around the coast. These slaves were then shipped under appalling conditions to the various colonies and re-sold into slavery, to work on plantations, in mines and as domestic servants. Shipping merchants then loaded their vessels with produce from the plantations and returned to Europe. The European merchants who provided the money for the trade, the shipowners and slave merchants, the African chiefs and the plantation owners made enormous wealth from these activities. Many European families and cities still enjoy the wealth so created.

Earlier in the century that Columbus made the first European landfall in the Americas, Portuguese and other European sailors were exploring the western coast of Africa, navigating around the Cape, and making first contacts with the well developed coastal settlements along the East African coast. These early expeditions led to the first trading links between Europe and Africa south of the Sahara, and in the case of East Africa to raiding and plunder. Trade largely consisted of European traders exchanging their cotton and woollen goods, brass and other metal goods, swords, various clothing and so on for African gold, ivory, pepper and fast-dyed cloth. A handful of men and women who were already slaves in their own lands were also delivered back to Europe, mostly to work as servants.

The impact on Africa was disastrous. It is calculated that in the centuries when slaving was widespread about 12 million living slaves were landed in the

A painting called *Scene on the Coast of Africa* by A. F. Biard

Americas, while probably a further 8 million died as a result of savage treatment or illness either in Africa or on the slave ships. They were the physically most able men and women, often well skilled in local crafts, and their loss was a great drain on these parts of Africa. The effects of violence and moral disintegration are hard to measure. Certainly, the new range of goods imported into Africa – including guns, alcohol and many textiles and metal items – did nothing to help the development of the people as a whole.

The colonisation of Africa

During the early and middle years of the nineteenth century, when slavery was gradually being abolished, European missionaries, explorers and traders were realising the enormous wealth that lay in the interior of the continent. European rulers, governments and businessmen wanted a share of this. In 1877 King Leopold of Belgium unashamedly said 'I mean to miss no chance to get my share of this magnificent African cake'! The European nations agreed to collaborate in this venture, and at the Berlin Conference of 1884–5 agreed on their spheres of influence. An English minister, Lord Salisbury, said at the time 'We have been engaged in drawing lines on maps where no white man's foot has ever trod. We have been giving away mountains and rivers to each other, only hindered by the small impediment that we never knew exactly where they were.'

Through a mixture of dubious treaties, threat of force or actual warfare, often under the pretext of putting down 'rebellions' of Africans who resisted the take-over of their lands, the Europeans occupied and ruled most of Africa by the turn of the century. The manner of the British methods were well exemplified by Cecil Rhodes, who had the double ambition of making a great deal of money from his activities in Africa and seeing the British flag in every country 'from the Cape to Cairo'. He died in 1902, but his dream was realised in 1915 when Britain took control of Germany's East African colony of Tanganyika.

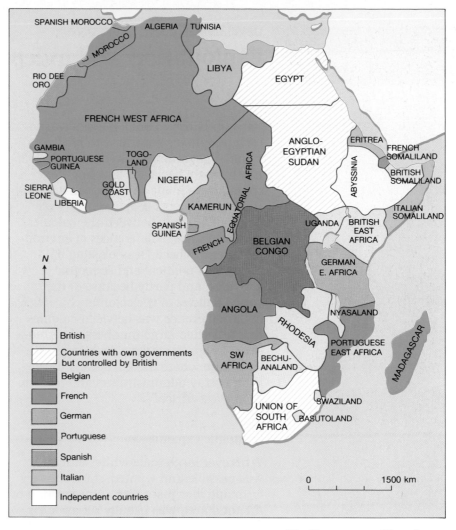

The European partition of Africa, 1914

The occupation and rule of these artificially created countries was possible through the use of military force by the European powers. Undoubtedly the motives of the many people involved – administrators, traders, settlers, soldiers, missionaries and so on – varied a great deal, and many did their best to help the African people. Many economic and social developments resulted from European involvement, most notably the creation of farms and plantations, mines, roads, railways, schools and hospitals. But there is no doubt that Europeans gained a great deal from their occupation, and many Africans feel that they lost far more than they gained.

1   a) Name the countries between the Cape and Cairo once ruled by or on behalf of Britain.
    b) Which two countries were independent in 1900?
2   List some of the gains and losses for the Africans of European colonisation.

Exercises on page 224
Cartoon views

Cecil Rhodes, striding from Cape Town to Cairo

83

4.3 Exploitation and independence

The aim of the colonial powers was to replace military control with civilian colonial government as soon as possible. All the main decisions affecting the colony were made by European administrators, not local people. There were many differences between the colonies, partly because of differences of environment, partly because of existing differences in political and economic activities, and partly because of the varied policies of the colonial powers. A major difference was between colonies that attracted large numbers of permanent European settlers, such as Kenya, and those where the Europeans were shorter-stay administrators and traders, as in West Africa.

Economic exploitation

Wherever large-scale white settlement was possible and wanted, the necessary farmland was just taken. A proportion of the total area was usually set aside for the local people as reserves, but it was often the poorest farmland and lacking in rich mineral deposits. Land needed for roads and railways, plantations and mines was also taken as required. A widespread aim was to develop export crops such as cocoa, cotton, tea and rubber, and mineral raw materials, to send to the controlling country. If local labour was not sufficient, migrant workers were brought in from other parts of the colony, or from other countries. Some attempts were made to provide basic education, health care, government and law and order. To pay for these developments, and for police and troops, taxes were raised from the local people.

The European officials were powerful, but few in number. These colonial administrators had to rely on local clerks, interpreters, police constables, servants and so on to govern the colonies. These intermediaries between the Europeans in charge and the local population wielded quite a lot of local influence, but had very little say in major decisions. Many white officials worked devotedly and faithfully, but were often caught between the colonial policy of making a profit from the land and its people and their wish to help local people by providing them with services and welfare.

The colonial powers regarded their colonies as overseas 'estates' from which to gain profits and as bases from which to protect their empires and trade routes. They were seen as important sources of food and raw materials for their factories. Cotton, for example, was often grown on a large scale on land that could have been more wisely used for local food. This cotton was sold by the colonial administrators at artificially low prices and shipped to the mother country. Here it was manufactured into cloth and clothing, much of which was shipped back to the colonies and sold at artificially high prices! Through the protection of such trade the European companies and countries made considerable profits, while the local peoples lost land, worked hard for low wages, and found their own manufactures and foods difficult to sell in the face of the flood of imports from the ruling country.

The two zones

In most colonies two distinct types of area could be found. One consisted of those places where export products – minerals and crops – were produced.

In 1868 diamonds were discovered in South Africa. This led to the first major migration of black labourers, mainly from Basutoland (modern Lesotho)

These were linked to the sea by railways and roads, and included small towns where Europeans lived. These were the favoured zones of colonial development. Much of the remainder of the country, often huge areas, was known only by the local people who lived there and the occasional white administrator. Sometimes a missionary station with a school room and basic medical care might be found. Such zones were governed as cheaply as possible, and the benefits of colonial rule were few. Their great value to the colonisers were as a source of labour for the farms, plantations, mines and factories. Massive temporary migration of labour led to rural areas being almost empty of young males for much of the year. This meant not only the break-up of families but also farming being made the additional responsibility of women. Underlying these economic practices was a widespread neglect of local peoples, and a failure to provide them with education or responsibilities for running their own lives.

It was small wonder that after many decades of such exploitation and subservience most of the local peoples wanted to be free of their colonial rulers. In the decades following the Second World War most of the colonies gained their independence, sometimes only after violent struggle. However, the lives of many people in the former colonies have hardly improved since independence, and most of these newly-independent countries are struggling with many problems. Part of the explanation may lie in their recent exploitation by European colonisers and the lack of preparation for self-rule.

This advertisement for Lipton's Teas appeared in *The Illustrated London News* in 1896, and shows scenes from Ceylon (now Sri Lanka)

1 **a)** From which British colonies were the minerals alumina, tin and copper obtained?
  b) From which were cocoa, tea, rubber and palm-oil obtained?
2 List any **a)** advantages **b)** disadvantages you can think of for the colonised countries as a result of their being colonised.
3 Which British colonies did not become or remain members of the Commonwealth?

Exercises on page 225
St Christopher-Nevis

The Commonwealth

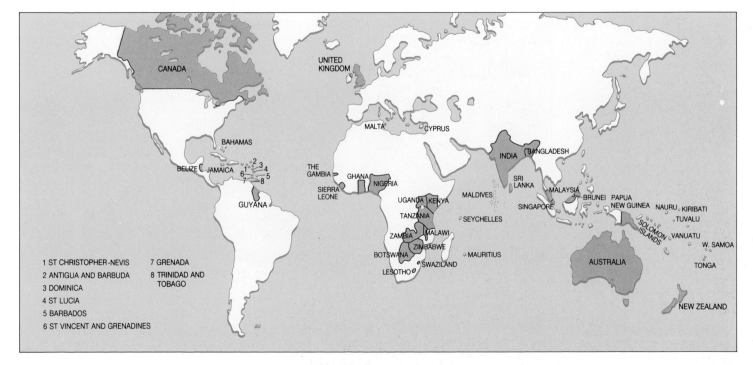

1 ST CHRISTOPHER-NEVIS
2 ANTIGUA AND BARBUDA
3 DOMINICA
4 ST LUCIA
5 BARBADOS
6 ST VINCENT AND GRENADINES
7 GRENADA
8 TRINIDAD AND TOBAGO

4.4 World trade

Copper smelting in Zambia. Zambia is heavily dependent on its exports of copper

Many Third World countries are very dependent on world trade. They sell products from their farms, mines and factories and use the resulting income to pay for goods imported into their country and for public services. There is nothing unusual in this, and all the countries in the so-called developed world, both those with market economies such as the United Kingdom, Australia and the United States, and those with centrally planned economies such as the USSR and China, do the same. The economic well-being of countries depends a great deal on the success or failure of this world trade. The most fortunate countries are those that can rely on earning a lot from their exports while needing to spend only a small amount on essential imports. Unfortunately very few Third World countries enjoy such a healthy balance of trade.

Dependency on a few commodities for export

The very poorest countries are almost unaffected by world trade, but we have seen how many Third World countries in colonial times were 'developed' to provide foodstuffs or industrial raw materials. Since independence they have continued to rely on the export of these basic or primary goods. The main problem is that many countries rely very heavily on just one or two export products. Some idea of the location of such countries can be gained from the map. When prices for these products are high then income gained is also high. But the demand and price for primary products is notoriously unstable. During the 1980s, for example, sugar, iron ore, crude fertilizer, copper and tin have all shown dramatic drops in price, and this has had disastrous effects on countries that rely on them as a major source of income. The prices for basic commodities are linked with general economic well-being, but are also controlled by large financial concerns through international commodity markets.

Countries import goods that they are unable to produce themselves, or can but at prices that cannot compete with those of imported goods. One of the consequences of using large areas of good quality farmland for export crops is that countries that were once self-sufficient in food production now have to import food. Most Third World countries have only a limited manufacturing sector, and they have a continuing need to import manufactured goods and fuels. During the 1970s the prices of many

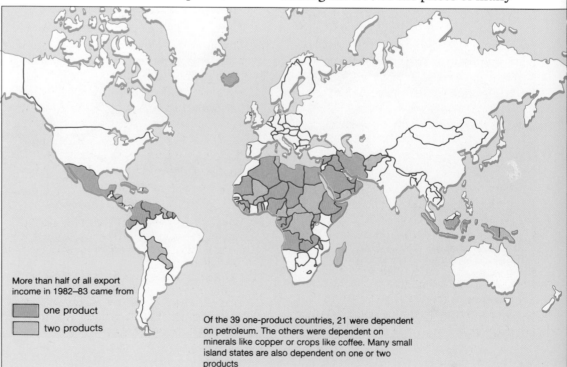

More than half of all export income in 1982–83 came from

■ one product
□ two products

Of the 39 one-product countries, 21 were dependent on petroleum. The others were dependent on minerals like copper or crops like coffee. Many small island states are also dependent on one or two products

Countries dependent on one or two exports

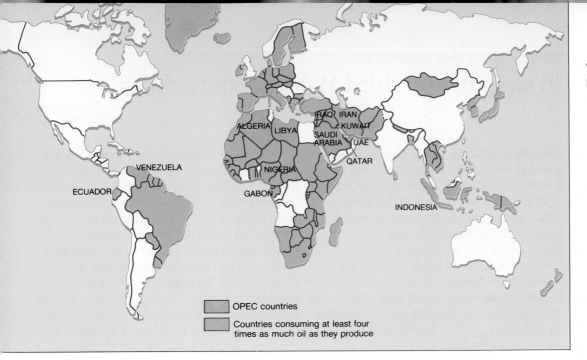

OPEC countries

Countries consuming at least four times as much oil as they produce

basic commodities fluctuated widely while the prices of manufactured goods were fairly stable but rising. Overall this has meant that the purchasing power of Third World exports has fallen back just as the costs of imports have risen.

Organisation of Petroleum Exporting Countries (OPEC)

One of the most remarkable changes in economic trade between the 1950s and the 1970s was based on an enormous increase in the demand for and production of oil. The demand was largely from the developed market economy and centrally planned countries while the producers included many Third World countries. For a while these oil producing countries experienced a marvellous boom, based on their ownership and control of one of the few things the more wealthy and powerful nations really needed. These countries formed themselves into the Organisation of Petroleum Exporting Countries (OPEC). Their new-found economic power enabled them to quadruple the price of oil during shortages arising from war in the Middle East. The resulting wealth was used to amass enormous personal fortunes, develop industries and provide a whole range of services in the developing countries. Some countries borrowed even larger sums from international banks to invest in further developments. By the mid-1980s, however, the oil producing countries were caught in the dilemma of over-production and reduced demand. As demand fell, so did prices, and the boom in this vital commodity seemed about to end. The Third World oil producing countries did well out of the boom, but others less fortunate had also to pay the high prices for their imported oil.

Newly-industrialising countries

One of the possible solutions to over-dependence on one or two export commodities would be to become more economically independent by manufacturing things within the country. During the 1970s and 1980s a few countries such as Brazil, South Korea, Taiwan and Singapore managed to make remarkable progress in industrialisation. Not only are they providing for their own needs, but are competing in world markets with the older manufacturing countries. As we shall see, such efforts have sometimes failed, and it is important to realise that there are many contrasts in the economic health of Third World countries.

1 Name four countries that are both heavily dependent on one or two export crops and also consuming at least four times as much oil as they produce.
2 **a**) From examples elsewhere in this book name other Third World countries that have important manufacturing industries. **b**) Why do not all Third World countries industrialise like South Korea (see page 126)?

Exercises on page 226
Zambia's dependence on copper

Transporting Third World commodities

The very nature of trade means that goods have to be transported from one place to another. In international trade this involves the movement of huge amounts of materials, some in liquid form, some in dry bulk, and some in small units and containers. Since so much Third World trade is with the countries of Europe, North America, the USSR, Japan and Australasia, this means the use of air or sea transport. The type of transport used is determined by the amount and value of the cargo – it would be enormously expensive to send large amounts of dry or liquid bulk goods such as iron ore or oil by air, for example.

Land transport to the ports

Most farms, plantations, oilfields and mines are some distance from ports used in international trade. From the beginning it was realised that there was little point in producing goods for export if they could not be got to these ports fairly quickly and cheaply. Consequently, during colonial times high priority was given to the building of roads and railways from the centres of production to the coast. Nowadays many loans and aid schemes for Third World countries are to develop transport links.

The difficulties of road and railway construction are often very great. Taking ores out of the Andes mountains in Bolivia or from mines in Zambia or Zaire in the centre of Africa, or plantation and farm crops out of the Highlands of Kenya, for example, present many problems. In some cases the building of roads and railways is complicated by very difficult terrain or severe and damaging weather. In other cases vast distances must be overcome. Sometimes the most pressing problem is the lack of modern and efficient lorries or railway engines and rolling stock. Some countries must additionally cope with the difficulty of being land-locked. They can only export to and import from distant countries if their neighbouring states give them road or rail access and allow aircraft to fly through their airspace. Good examples of this land-locked situation exist in Southern Africa where Zambia, Zimbabwe, Botswana, Malawi, Lesotho and Swaziland all have to rely on the goodwill and transport systems of neighbouring countries. When relations are good then all is well and everyone benefits from the trade. But if any sort of conflict arises then the economic trade of the land-locked state can be virtually stopped. Similar conditions exist when the commodity is oil and transport to the exporting ports is by pipelines that cross neighbouring countries.

The map shows the major railways and ports in Southern Africa. Notice the land-locked location of some countries. The photograph shows maize exports from Zimbabwe being loaded onto ships at Durban docks, South Africa. The maize is carried from Zimbabwe by train, and is destined for West Africa. The dockers are migrants

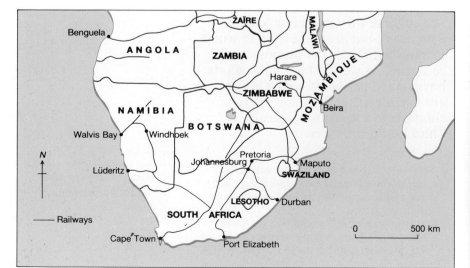

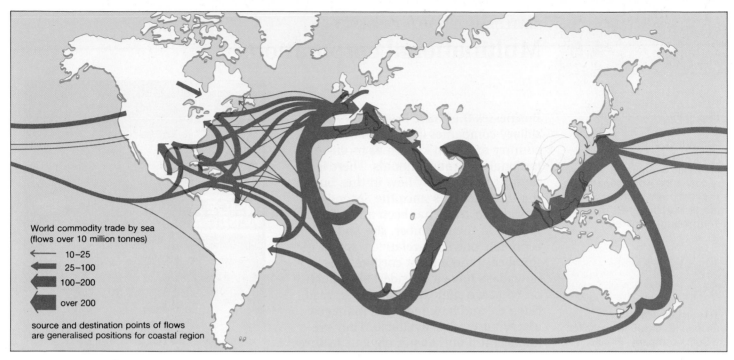

World commodity trade by sea
(flows over 10 million tonnes)

→ 10–25
→ 25–100
→ 100–200
→ over 200

source and destination points of flows
are generalised positions for coastal region

World sea routes

Ocean-going trade

Some idea of the volume and pattern of ocean-going trade can be seen on the map. Even though actual ports are not shown, nor individual countries, it is clear that there is considerable trade between Third World countries and others. In the past most ocean-going trade was carried by general cargo ships, but many changes have occurred in recent decades. Ships have become much bigger to cut the costs of transport. Specialised oil-tankers and bulk coal and ore carriers and container ships have been developed to make the transport of these various goods much quicker and cheaper, and to reduce the time that ships wait in port to be loaded and unloaded. The growth of international trade has gone on hand in hand with the growth of the world's cargo shipping fleets.

The ownership and country of registration of ships gives a false impression of the importance of trade to various countries of the world. Liberia and Panama, for example, are amongst the leading nations for shipping numbers and tonnages registered under their flags, yet they are relatively poor countries. Registering ships in this way is a means of reducing running costs and avoiding many regulations insisted on by many other countries. But Third World countries have increasingly contributed to world shipping and to world trade. With the exception of the very poorest their very survival depends on transport and trade. As world trade as a whole declined during the 1980s, so did the transport of products and basic commodities. Many ships are laid-up and temporarily out of use in Norwegian fiords and Greek inlets and other sheltered harbours around the world.

1   What is the weakness of the map above in showing **a)** the importance of ocean trade to Third World countries and **b)** important route patterns?
2   From other sections in this book, or from other sources, name and locate important Third World exporting ports. Give two from each continent.
3   Name four land-locked countries that depend on the export of primary products. For each name the neighbouring country and port that it could use if agreement could be reached with that country.

Exercises on page 227
The Trans-Gabon railway line

A seaport with modern facilities is an important asset to any developing country with a coastline. This is Lagos, Nigeria

4.6 Multinational corporations

(Right) the cover of *New Internationalist*, a magazine concerned with Third World issues, showing goods produced by Unilever. Unilever is one of the world's major multinationals, with profits larger than the GNPs of most Third World countries

This Singapore factory assembles cassette decks for a Dutch company, Philips, for export to Britain

Businesses that have branches or subsidiary companies in more than one country are often known as multinationals or transnationals. There is nothing particularly new in this, but in the three decades since the Second World War there has been an enormous growth in their number, size and economic power. A relatively small number of these giant corporations nowadays have a considerable influence on what we can eat, drink, wear, read, listen to, and how money is managed and what work is available. They are important if not always obvious features in the lives of most people.

Features of multinational corporations

A vital aim of businesses is to make profit for the owners and investors. If there are benefits for other people, or the nation as a whole, so much the better, but making profits is the primary goal. Multinational companies have come into existence and grown to such importance because large scale seems to offer the best chance of security and larger profits. Through take-overs and mergers smaller companies get absorbed into more powerful ones, and so the process of growth continues. As can be seen from the extract, the trend is towards the formation of even larger global corporations.

Some multinationals are of very great size in terms of financial turn-over, profits, number of employees and variety of products. The largest, such as British Petroleum and Unilever, General Motors, IBM and Hitachi, have a bigger turnover than dozens of countries. They are registered in the country of the parent company, and all major decisions that affect their overseas operations rest with the owners and managers at the centre of the organisation.

Another feature of multinationals is their diversity. This may be 'horizontal', meaning they are involved in a great range of products or services, or 'vertical', meaning that a company controls all or most of the steps of activity from production of raw materials, through processing, transport, management, advertising and sales.

Multinationals and the Third World

Most of the investment, trade and profits of multinational companies relate to countries outside the Third World, but multinationals do have a very powerful influence for good and bad on less developed countries. It is calculated that about one-third of the economic output and trade of Third World countries is controlled directly or indirectly by multinationals, and many governments encourage them to set up business within their countries. They usually become involved in the extraction of minerals, the large-scale production of foods and vegetable raw materials on plantations, and in manufacturing. Manufacturing is often concentrated in free-trade zones where there are special tax and other advantages. Goods manufactured in these free-trade zones may be largely for export, or to avoid import levies if they are to be sold within the country. Investment in mines and plantations is widespread throughout the Third World wherever mineral deposits are found or crops can be grown. Manufacturing is more restricted to countries such as those in OPEC or newly-industrialising nations such as Singapore, Korea and Brazil.

These developments are supposed to benefit Third World countries in a number of ways. They provide employment, while the wages mean there is spending power to boost local production of food and goods. The multinational companies bring extra work to the area, buying all sorts of raw materials, goods for their mines and factories, and many services. Above all a part of their profits are paid in tax to the state. So both the multinational company and the people of the country with the subsidiary company or branch should benefit.

In practice it does not always work out this way. Wages may be relatively high, but one of the major attractions for the multinational company is cheap and well-ordered labour. Women are often employed, since they can be paid less and are likely to be less unionised. Many raw materials and products are not bought locally, but are brought in from other parts of the company in other countries. Profits are sometimes kept artificially low – and hence tax payments are also low – through financial arrangements with other parts of the company outside the country. Last but not least the owners and managers of the parent company can make decisions to close down works, or move their operations, and so cause unemployment,

Why global advertising ... 'is it'

For multinational manufacturers of brand name consumer products the market is now the world. One of their major weapons lies in multi-million pound marketing budgets, and the trend is towards global brands.

The *global* corporation operates as if the entire world were a single market, and it sells the same products in the same way everywhere – at relatively low cost. Multinationals, on the other hand, operate in several countries and adjust their products and practices in each – at relatively high cost.

Thus Coke 'is it' all over the world.

if the company as a whole needs it to increase profits. Such 'rationalisation' of operations takes place world-wide, but its impact can be most devastating in poorer countries.

> Exercises on page 228
> Multinational mining companies in Namibia

1 What are the products and where is the home country of the following multinationals: Mobil, General Motors, IBM, Hitachi, Nestlé, BAT, BASF?
2 List some of the products and countries mentioned in the RTZ report. What advantages of multinational operation to the Third World countries are mentioned in the report? What happened in Zimbabwe?

Extract from RTZ annual report

The Rio Tinto-Zinc Corporation PLC

'Our group is fortunately well spread both geographically and by product'

Results
Our investments in Australia and Canada have been through a difficult time, the recession hitting them hard.

In Spain, Rio Tinto Minera showed a loss reflecting high operating costs and lower metal prices. In Zimbabwe the Empress mine regrettably, but with Government agreement, has now been closed and RTZ has given Z$750,000 to assist with a relocation and rehabilitation programme for former employees.

The good news is that the Borax mine in California and both Rössing Uranium and Palabora in Southern Africa made satisfactory profits.

Mining and the Third World
Large scale mines such as Bougainville in Papua New Guinea are unlikely to be developed over the next few years. It was decided not to proceed with the Cerro Colorado development in Panama. This demonstrates the problems involved for Third World countries in developing new resources. International mining companies make a major contribution by bringing together financial, technical and managerial expertise that these countries can seldom obtain elsewhere.

Their involvement greatly increases the likely success of new mineral development with substantial economic benefits to the host countries. Among these are higher overseas earnings, greater employment opportunities and an injection of money into the economy. For local people there is the prospect of higher standards of living and greater opportunities for training, acquiring new skills and a wider education.'

4.7 Third World imports

We have seen that some people claim that the industrialised and more economically developed nations contribute to the poverty and lack of development in the Third World by taking their resources at unfair prices and using their labour for low rewards. It is also claimed that what is sold to Third World countries often takes advantage of their lack of knowledge and their pressing needs. Third World countries, it is argued, are exploited as much by what they are sold as by what is bought from them.

Commodity 'dumping grounds'

The 'dumping' of products occurs when manufacturers cannot sell their products in their own country because they have been declared unsafe or hazardous and so export them to countries where their sale is still permitted. A wide range of products are involved, from dangerous electrical equipment, toys and clothes to certain brands of foods. Above all many drugs, pesticides and chemicals used in agriculture and manufacturing that are banned by governments in the industrialised world are sold to people in Third World countries. Proven cases include British-made sweetened conden-

Baby milk

'Breast is Best' campaigners have given Nestle, the Swiss-based food conglomerate, an almost clean bill of health. They have switched their attention to other US, European and Japanese companies who are still busy promoting baby milk in the Third World.

Manufacturers persuade mothers to abandon breast feeding and use powdered milk instead. But in developing countries where money, clean water and fuel are all scarce, bottle-fed babies often suffer from diarrhoea. Some die through dehydration.

sed milk sold in South-East Asia as 'excellent for infant feeding' when in fact it was highly unsuitable, causing obesity and deficiency in some nutrients. It is claimed that virtually every pesticide banned in the USA has been exported to other countries. For years dried milk powder was aggressively sold through powerful advertising to mothers in Third World countries as an alternative to breast feeding when it was known that due to lack of clean water and sterilising facilities the practice was dangerous.

Manufacturers and suppliers are able to dump goods in this way for several reasons. Third World countries usually lack proper means of controlling imports due to a lack of trained people and laboratory equipment for testing products. Government negligence, corrupt officials who let things through for their own personal profit, and the sheer ignorance or illiteracy of the users all contribute. Much depends on the truth of information given by instructions and

In Bandung ...

In Bandung in Indonesia, Siti visits the market. She picks up a bottle of herbicide, containing 2,4,5–T, and asks for an anti-diarrhoeal drug for her son Bambung ... she gets Entero-Vioform.

Siti didn't know that 2,4,5–T is a severely restricted herbicide in the West. She didn't know about its suspected mutagenic properties. She didn't know that Entero-Vioform contained clioquinol which had caused SMON nerve disease killing hundreds and paralysing thousands in Japan. Siti didn't know she was being dumped on.

Cigarette smoking is on the decline in the West. Tobacco companies see the Third World as an untapped market where they can make up for the loss in sales elsewhere. This advertisement is in Jakarta, Indonesia

advertisements, and people in the Third World are no different from others in being persuaded by advertisements and claims. Few realise that they may be buying second-rate rejects, goods actually prohibited in the country of origin or that they should only be used in very special circumstances.

Much depends on the honesty and lack of greed of the producers and suppliers. If they are prepared or even eager to sell products that are proscribed as dangerous in their own country they can change its name or make small variations in the ingredients and sell it as a non-proscribed product. They may produce for export only, avoiding strict testing, or manufacture or put together the ingredients in the country of sale. It is difficult to assess the extent of dumping of dangerous or inappropriate products in the Third World, because figures naturally are not available, but it has been called the 'corporate crime of the century'.

Arms sales to the Third World

Manufacturing weapons of all descriptions is one of the major industries of the world, and spending on arms is greater than on health or education. Most spending on arms is within the so-called developed world, and major manufacturing and selling countries are the USA, the USSR, Britain and France. Many weapons, ranging from personal firearms to modern tanks, airplanes and rockets are bought by Third World

East trebles its arms sales to Third World

The Soviet bloc's arms sales to non-communist Third World countries have virtually trebled since 1975, a NATO study shows. However, its development aid to these countries declined.

The study says arms sales to developing states are probably Moscow's second biggest source of hard currency.

The report gives no comparative figures for Western arms sales to the Third World, although independent institutes say the United States, France and Britain are major suppliers.

countries. Governments claim they are necessary for national defence against attack from other nations, or to overcome and control anti-government opposition from within their countries. Guerrilla movements purchase huge amounts of armaments through illegal arms dealers. Everyone engaged in conflict, or anxious about national defence or even wanting modern weapons as symbols of power, will argue that this is money necessarily spent. Nevertheless these purchases add greatly to the import bills of countries that are often desperately short of resources and frequently heavily in debt. The purchase and sale of arms is different from the purchase and sale of dangerous, hazardous and inappropriate products. The former may be an unfortunate necessity in a world where the activities of nations and groups is still dominated by suspicion, fear and conflict. The latter can only be justified by those who say that the profit of the company and traders is the only thing that counts. In both cases the trade contributes to the delay in the economic and social development of some people.

1 What are some of the dangers in the use of dried milk, tobacco, asbestos and pesticides in Third World countries? Would it still count as 'dumping' if the purchasers and users knew of the dangers?
2 Do you think it acceptable or not for companies to advertise and sell goods that they know are hazardous without making it absolutely clear?
3 What are the advantages and disadvantages of manufacturers and countries in the industrialised world selling armaments to Third World countries? What are the advantages and disadvantages to the Third World countries?

In spite of many restrictions in their day-to-day life, Iranian women are still taught to use high-powered weapons like this rocket-propelled grenade launcher

Exercises on page 229
Dumping

4.8 Third World loans and debts

All countries of the world, even the poorest, are part of the world global economy. This involves the buying and selling, importing and exporting, of many products and services. It also means that all are involved in the international flow of money, in borrowing and lending and receiving and repaying. Generally speaking the countries of the Third World are the borrowers and repayers and the lenders are countries of the more developed world.

Third World borrowing

Countries and enterprises borrow money for different purposes. It may be to develop the economy by opening mines, extending farms and plantations and building factories, or by constructing the power stations and transport links needed for the economy to flourish. They may borrow in order to build schools, clinics or hospitals and to develop services concerned with the quality of life of the people. Sometimes borrowing is for arms and weapons to support the government and its administration. Different governments give different priorities to their needs. Being an independent country enables these decisions to be made by the people or their government rather than by some other country. But loans have to be repaid, and the lending countries, agencies and

Debt interest as percentage of exports

		1985	1984
1	Argentina	52	56
2	Chile	46	44
3	Brazil	41	38
4	Mexico	33	37
5	Peru	31	33
6	Philippines	28	29
7	Ecuador	24	30
8	Venezuela	18	16
9	Yugoslavia	15	16
10	Nigeria	13	11

The table shows the percentage of export income that was used to pay debt interest.

The danger is when export earnings are less than interest payments!

The leaders of some of the world's richest countries attended a summit meeting in Toronto, Canada, in 1988, and one of the things they discussed was the problem of Third World debt

Official and private debts

Less developed countries are involved in two types of debt – official and private.

Official debt is owed by governments and enterprises to non-bank financial agencies, normally in the developed world. These include governments and international agencies such as the IMF and World Bank.

Private debt is owed by governments and enterprises to private international banks. Often more than one international bank is involved in the loan.

Loans from official sources are normally medium or long-term. Loans from private sources are more likely to be short or medium-term.

banks are interested in how the money will be used and how certain they can be in having it repaid. Governments are more likely to lend to countries with similar political goals or that they think they can influence. The basic aim of banks is to make a profit and extend their business, so they will favour lending to countries that can guarantee repayments with whatever interest is asked. International agencies can afford to be less limited and constrained about whom they lend money to, but need to be assured that it is being put to what they, the lenders, think are good purposes.

During the 1970s there was a general economic prosperity and a growth of world trade. Lots of money was available from oil-rich states, through international banks, for loans to whoever wanted them. Income from the sale of commodities was relatively high, and Third World countries could afford to borrow with confident expectations of being able to repay their loans. The poorest of Third World countries borrowed to try and start their economic growth and social development, while others such as Mexico and Brazil in Latin America and South Korea and Taiwan in South-East Asia borrowed to allow further industrialisation.

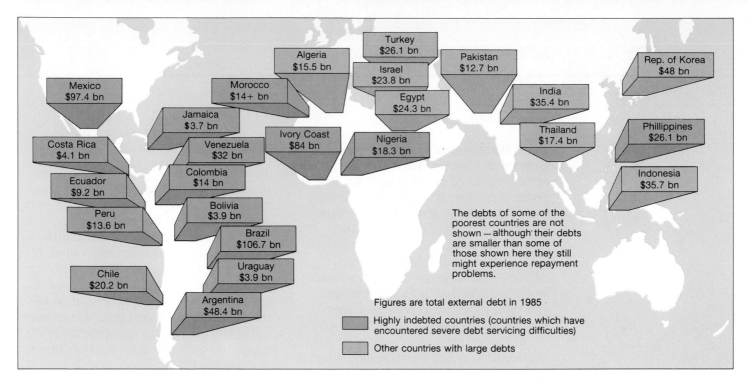

Mexico $97.4 bn

Costa Rica $4.1 bn

Ecuador $9.2 bn

Peru $13.6 bn

Chile $20.2 bn

Jamaica $3.7 bn

Venezuela $32 bn

Colombia $14 bn

Bolivia $3.9 bn

Brazil $106.7 bn

Uraguay $3.9 bn

Argentina $48.4 bn

Morocco $14+ bn

Algeria $15.5 bn

Ivory Coast $84 bn

Turkey $26.1 bn

Israel $23.8 bn

Egypt $24.3 bn

Nigeria $18.3 bn

Pakistan $12.7 bn

India $35.4 bn

Thailand $17.4 bn

Rep. of Korea $48 bn

Phillippines $26.1 bn

Indonesia $35.7 bn

The debts of some of the poorest countries are not shown — although their debts are smaller than some of those shown here they still might experience repayment problems.

Figures are total external debt in 1985

☐ Highly indebted countries (countries which have encountered severe debt servicing difficulties)

☐ Other countries with large debts

The growth of world debt

At the end of the 1970s and during the first half of the 1980s the situation changed dramatically. For a variety of complicated reasons world trade and economic growth slowed down and the more developed nations began to suffer severe economic problems and high unemployment. There was a fall in demand for raw material commodities from Third World countries, and the industrialised nations discouraged or prevented the importation of manufactured goods to protect their own manufacturers. By the middle of the decade the price of oil also fell dramatically.

At the same time the industrialised nations and their banks increased the cost of borrowing very severely by raising interest rates. These two simultaneous events – the decline in income from exports and the increase in debt repayments – meant that for reasons beyond their control some Third World countries were sliding close to bankruptcy. Large proportions of their foreign earnings were being used to pay the interest on their loans, not to repay the loans themselves! There were only a few solutions. One was to spread out the repayments over a longer period, if it could be agreed. Another was to borrow even more to enable the older debts to be repaid. But these would be short-term loans charging much higher interest.

Others were to refuse to make any repayments or to go bankrupt, both of which would cause chaos within the country and in the international financial markets. The mid-1980s was a period of potential international debt crisis.

Many countries applied to the International Monetary Fund (IMF) and World Bank for new or rescheduled loans. They were usually given, but only on strict terms. Almost inevitably these meant cutting public spending on health, schools, infrastructure, economic developments, armaments and so on; ending food subsidies and price controls; and devaluing money. All these meant harsher conditions for the population and unpopularity for the government. The strict terms made sense to the lenders, but many Third World countries felt they were once again under the control of wealthier nations, either individually or collectively.

1 From the map give the names of two countries with debts that are a) amongst the poorest of Third World countries b) Third World middle-income countries having repayment difficulties c) Third World countries in South-East Asia without repayment problems.

2 Imagine you are a representative of either the IMF or the government of a lending country. Describe your thoughts about conditions being laid down for a loan to be granted. What is for and what is against the loan?

The world's major 'debtor nations'

Exercises on page 230
The debt problem

This cartoon neatly sums up what most people feel about the relationship between the 'North' and the 'South'

4.9 Aid

Aid is the giving of resources from one country to another either absolutely free or very cheaply. Development aid is that given to countries to help them develop their economy or services and so raise the quality of life of the people. It is usually distinguished from emergency aid, following a disaster, and from military aid, although often they cannot be easily identified. It includes people and their expertise as well as money, food and other resources.

Aid given from one country directly to another is called bilateral aid. When it is offered as a contribution to international organisations such as the World Bank, and redistributed to a number of countries, it is known as multilateral aid. About 60 per cent of British aid is spent bilaterally. Some aid is given for special developments such as road building, providing water supplies or developing health and school facilities. This is known as project aid, and gives the donor some control over the use of the money and some knowledge of the progress or success of the project. Programme aid is for purposes other than special projects, such as helping countries out of balance of payments difficulties for example. This gives the receiving country more choice in how the aid is used.

Who gives aid?

The diagram shows who gave development aid during the early 1980s. The first distinction is between governments, who gave by far the most, and non-governmental agencies whose contributions were smaller but often vital. It is interesting to compare the amounts given in overseas aid with other spending. In 1983 Mr Clausen, President of the World Bank, stated that 'more than $600 billion is being spent annually on weapons compared with $125 billion for development.'

The motives for governments giving aid are numerous and varied, and their contributions are better measured when given as a percentage of their gross national product than as an absolute figure. The United Nations has suggested a target of 0.7 per cent of GNP. The average Western aid in the early 1980s was 0.35 per cent. Britain contributed about 0.44 per cent while Norway, Sweden and the Netherlands exceeded the UN target.

Donor countries often prefer aid to be 'tied' as it is the type of arrangement which offers them most benefits. Untied aid means that the receiving country can spend the money for development work anywhere it likes on the open market. Tied aid means that any goods and services needed during the uses of

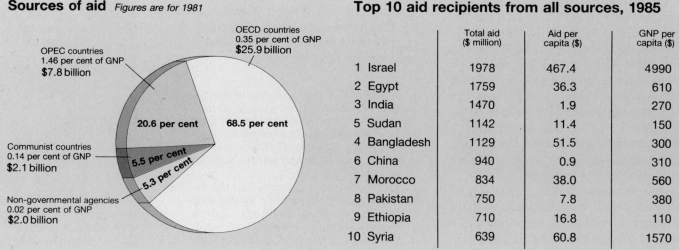

Sources of aid Figures are for 1981

OECD countries
0.35 per cent of GNP
$25.9 billion

OPEC countries
1.46 per cent of GNP
$7.8 billion

Communist countries
0.14 per cent of GNP
$2.1 billion

Non-governmental agencies
0.02 per cent of GNP
$2.0 billion

20.6 per cent
68.5 per cent
5.5 per cent
5.3 per cent

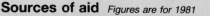

Top 10 aid recipients from all sources, 1985

		Total aid ($ million)	Aid per capita ($)	GNP per capita ($)
1	Israel	1978	467.4	4990
2	Egypt	1759	36.3	610
3	India	1470	1.9	270
5	Sudan	1142	11.4	150
4	Bangladesh	1129	51.5	300
6	China	940	0.9	310
7	Morocco	834	38.0	560
8	Pakistan	750	7.8	380
9	Ethiopia	710	16.8	110
10	Syria	639	60.8	1570

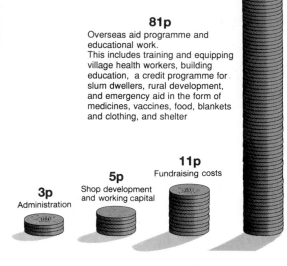

How OXFAM used every £1 in 1985–86

81p
Overseas aid programme and educational work.
This includes training and equipping village health workers, building education, a credit programme for slum dwellers, rural development, and emergency aid in the form of medicines, vaccines, food, blankets and clothing, and shelter

11p
Fundraising costs

5p
Shop development and working capital

3p
Administration

the aid must be obtained from the donor country. This means that the donor countries can themselves benefit from the aid they give in terms of employment for their own workforce and sales of machinery, equipment and services. Other countries benefit by giving aid only to countries that have similar political views as their own, or who can help them in international affairs. This political directing of bilateral aid is common amongst the communist, OPEC and Organisation for Economic Cooperation and Development (OECD) countries (the Western industrialised countries). People are also sometimes surprised by who receives development aid, and how this relates to their GNP per capita (see the table).

Non-governmental agencies

Although the amount of aid given by non-governmental agencies is very small when compared with official aid, it often provides vital relief in times of crisis and help where governmental aid is lacking. The agencies such as Oxfam, War on Want, Christian Aid, Save the Children Fund and Voluntary Services Overseas tend to operate through voluntary groups in the receiving countries rather than through governments. They are committed in most cases to working with 'the poorest of the poor'. They are largely funded through public donations,

and much effort goes into fund-raising. In Britain the activities of aid agencies are limited so long as they wish to be charities (and so save money through exemption from certain taxes). Laws relating to charities prohibit 'political activity', and therefore some agencies have special non-charitable sections to take on controversial campaigns that they feel are important. The problem is that people frequently disagree on what is 'political'. Some idea of the range of activities of a voluntary aid agency is given in the details about Oxfam's programme.

1   **a)** Name two countries from the OECD, OPEC and communist groups. **b)** Suggest one Third World country likely to get bilateral aid from Britain, France, the USSR, the USA and Saudi Arabia, saying why in each case.
2   Rank the top ten recipients of aid from all sources according to aid per capita. Is there any relationship to their GNP per capita? Suggest major donor countries in each case, saying why you think they give aid to that particular country.

Exercises on page 231
OECD aid, and British aid

Road building in Kenya. Some people argue that aid money is better spent on 100 miles of simple 'feeder' road like this than on 10 miles of tarred road

4.10 Does aid work?

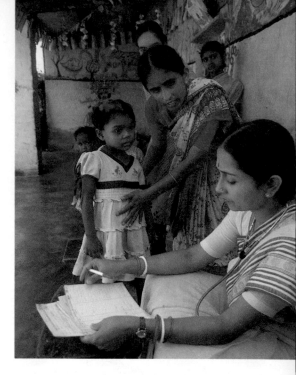

(Right) not all aid is in the shape of food or materials. Here Indian Save The Children Fund workers give medical care

At first sight it seems that development aid is worthwhile and important, and that more of it could only be a good thing. The influential Brandt Report *North–South: A Programme for Survival*, published in 1980, and its successor *Common Crisis*, argued strongly for an increase in aid to Third World countries. The reports claimed that not only did the richer and more powerful nations have a moral responsibility to help the less fortunate, but that development in the Third World was essential for the world as a whole. The First and Second Worlds of the North would benefit from the aid they gave both immediately and in the longer term. There are, however, some influential people who strongly criticise development aid programmes. Sometimes this is on the grounds that aid is wrong in principle. Other criticism is not so much that it is wrong in principle, but that it does not benefit the people who most need it – in practice it is used very badly. Some of the arguments against aid can be seen in the extract by Lord Bauer and Professor Yamey. The UK government and the many people who contribute to non-government agencies clearly think aid is worthwhile, even if it does need to be more effective.

This table, based on a World Bank report, suggests that World Bank aid aimed at the poorest people has not always been very successful

How much aid goes to the poorest people?

Reliable information about the impact of aid projects on the poorest people is difficult to obtain. A recent World Bank report, howerver, gives a fairly realistic assessment of the experimental 'poverty oriented' projects supported by the World Bank since the mid–1970s. Here are the report's main findings:

World Bank lending to 'poverty oriented' projects, mid 1970s–1981

Sector	Percentage of World Bank lending	Benefits for poorest groups
Rural development	15.3	Few direct benefits for the landless, for tenants and for near-landless farmers
Urban poverty projects	3.4	No benefits for poorest 20 per cent of the population
Small-scale, non-farm enterprises	1.7	No information on distribution of benefits
Water and sanitation	6.7	Only 30 per cent of benefits go to the poorest 40 per cent in urban areas only
Education	1.6	Less benefit to the poor than to better-off families. (Children of poor families work full-time)
Population health and nutrition	0.3	Not yet fully evaluated
Total	29.0	

Making aid more effective – one viewpoint

An article in a 1983 issue of *New Internationalist* proposed what it called 'Six Rules for Real Aid'. The first of these was to 'aim at the poorest'. It was claimed that aid directed at promoting overall economic growth by building power stations, copper refineries, steel-works, large plantations and so on is an ineffective way of helping the poor of any country. The predicted 'trickle-down' benefits of more employment, higher wages and better living standards for all do not happen. Aid should instead be aimed at landless farmers, share-croppers, tenant farmers and the urban jobless that collectively make up the poorest 40 per cent or so of populations in the Third World. This would mean more aid to provide water supplies, sanitation, credit for small farmers, basic education, primary health care and so on. It would probably mean less tied aid.

The second rule was to 'mobilise the poor'. They should be encouraged to get together to demand decent wages and working conditions and to break free of unscrupulous money-lenders, landlords and corrupt government officials who take aid money for their own gain. Rule three, 'fit aid to countries', suggested that aid should be given primarily to those governments, irrespective of their political views, that are genuinely committed to raising the living standards of the poor. Aid should be withheld from corrupt and inefficient governments where officials and powerful landowners or manufacturers would siphon off much for themselves. The fourth rule, 'rebuild the aid machine', argued that more money should be given directly to local voluntary groups, unions and co-operatives and peasant movements where the real 'cutting edge' of development occurs. A new international agency, independent of government and United Nations control, should also be created to handle donations from governments.

Rule five was to 'abolish sham aid', whereby funds are creamed off government aid budgets to subsidise those firms trying to win contracts to supply steelworks, refineries, aircraft and so on to Third World countries. In effect the development aid is used to subsidise manufacturers and businesses in the donor country that want to export. Rule six, 'have an independent audit', stressed that official aid is the taxpayers' money, and most people expect it to eventually benefit the poor, not the already well-off. The need is for an independent body to look carefully at how aid is used. That would mean, of course, having some agreed indicators of what successful aid really means.

The problem of deciding when aid has been successful has not been solved. Some people would measure it in terms of the rate of economic growth and the GNP, levels of financial reserves and debts, or the political views of the government in power. Others might be more concerned with the direct economic or political benefits they got from the aid, and be less concerned about people in the receiving country. Yet others would give emphasis to improvements in infant mortality rates, calorie intake, availability of safe drinking water, rising life expectancy and decrease in internal conflicts. Different people in the donor nations, and non-government agencies, have different measures according to their own state of well-being, beliefs and values. What is clear is that after decades of official aid there are still hundreds of millions of very poor people in the world, and that aid all too often seems hardly to touch them.

The arguments for and against aid put forward by experts. The first extract (top) was written by economists Peter Bauer and Basil Yamey, the second (bottom) was written by Timothy Raison MP, when he was Minister for Overseas Development, in reply

Why we should close our purse to the Third World

The argument that the rich countries of the North should give more financial aid to the poor countries of the South is familiar.

But experience has shown that:
- Aid cannot significantly promote Third World development.
- Aid does not relieve poverty in the Third World.
- Aid does not promote world peace or make friends for the West.
- Aid is neither appropriate nor necessary for relieving unemployment in the West.
- Aid is neither appropriate or necessary for solving the so-called international financial crisis.

Aid is not necessary for development. Many Third World countries have progressed rapidly, have even been transformed, without aid.

In reality, much aid goes to support projects and programmes which do not promote development and, going to governments, is apt to have serious adverse effects. It enables governments to introduce and persist with policies which are wasteful or which retard economic progress and discourage productive activities.

Aid has done little or nothing to help the poorest people in aid-receiving countries. Governments committed to improving the material standards of life of the poor are rare in the Third World. How do the poor, or for that matter the bulk of the population, benefit from the construction of new capitals (about $20 billion have been budgeted for Abuja, Nigeria), the operation of loss-making international airlines or vast prestige projects?

The wise use of an aid programme can lay the corner stone of development

The article by Lord Bauer and Professor Yamey deserves a response.

I agree with a number of points. Aid can in some instances be harmful to the recipient country.

Nevertheless, aid has an important role to play in helping to promote social and economic development in many developing countries. My job is to ensure that UK aid is used as constructively and as effectively as pos-

sible. We aim to do this by ensuring that projects are well-planned and effectively implemented and by helping to persuade governments to adopt more appropriate policies and assist in carrying them out.

It is unreasonable to write aid off as irrelevant or harmful. There is a great deal of good that properly managed aid can do.

1 List the main points made in the extract against giving aid. What are some of the counter-arguments?
2 In reading the 'Six Rules for Real Aid' what might be the objections of some to a) helping the poor to collaborate b) giving aid to countries that give priority to social rather than economic goals c) giving official aid?

Exercises on page 232
More views about aid

4.11 Superpower influences in the Third World

The bitter struggle between the American army and the communists in Vietnam lasted for 10 years and resulted in over a million deaths, the large majority of them Vietnamese (below). In 1988 the Soviet army began to withdraw from Afghanistan after a period of occupation (bottom)

Third World countries are not only part of the world financial and trade system, but are also involved in the manœuvrings and struggles of the major world powers. Some conflicts are based on religious differences or long-lasting border disputes over national claims to territory. The major threat to world peace in present times, however, lies in the struggle for supremacy between views favouring the capitalist market economy and views favouring the communist centrally planned type of economy. The former is represented by the United States of America and her allies, a grouping often referred to as the Western World. Their aims are to practice and defend democracy as a political system and control by market forces as the basis of economic organisation. The latter communist view is represented by the USSR and her allies who prefer a system of one-party socialism and an economy planned and controlled by central government. There are many different versions of capitalism and communism, of course, even to the extent of finding strong anti-American and anti-Soviet Union feelings in the respective blocs. Nevertheless the political and economic views represented by these two 'Superpowers' dominate world affairs and certainly have an impact on Third World countries.

Problems arise because neither side trusts the other. The communist world believes that the aid of Western governments is to perpetually organise world affairs for the good of the privileged, and that they will undermine communist countries whenever they can. The Western countries fear that the communist countries aim to slowly but surely ensure that all others will one day have communist governments, either through revolution or by direct invasion. So both sides spend vast amounts of money on defence (which the other side usually sees as a threat to attack) and on efforts to influence countries that do not belong to one or other group (they will try to disrupt those that belong to the other bloc!).

Many Third World countries do not want to be controlled by either bloc, even if their sympathies clearly lie in favour of one or the other. In 1986 there were 100 countries describing themselves as 'non-aligned', and in that year they held their international conference in Harare in Zimbabwe. In practice they tend to favour one of the two major blocs, and will be most involved in trade and aid with countries from that bloc. But they often manage to

NATO and Warsaw Pact countries

Key:
- North Atlantic Treaty Organisation
- Warsaw Pact Countries
- Countries mentioned in the text

make the most of competition between the superpowers by dealing with both. The superpowers and their allies try to win support in a variety of ways.

Methods of influence and interference

Superpower involvement is most obvious when it is of a military nature, either by providing military support (weapons and training for example) or by an invasion and direct participation in any fighting. One of the most devastating conflicts since the Second World War was in Vietnam. The Americans wanted to prevent communist North Vietnam taking over South Vietnam, and so they helped the anti-communist government of the latter. At first this was by giving material support and advice, but later by direct involvement in fighting the war. The North Vietnamese, with enormous resource backing from China and other communist countries, finally won the war, and a united Vietnam now has a communist government.

The USA and her Western allies support governments all over the world in attempts to prevent such take-overs and the spread of communism. Governments in El Salvador, South Korea, the Philippines and Pakistan, for example, have all received such support. In other countries with communist regimes, such as Nicaragua, Angola and Mozambique, Western policy is to destabilise the country and bring about the overthrow of existing communist governments.

The USSR and her allies do very

much the same. In 1979 the Soviet Union sent large numbers of troops into Afghanistan to support a dubiously elected communist leader and government, and became involved in a long and bitter guerrilla war against opponents of the regime. They support revolutionary groups in countries all over the world, and are the main providers of economic and political help to countries such as Cuba, Nicaragua and Ethiopia that have socialist principles and communist sympathies. Even when the USA and Soviet Union do not get directly involved in Third World countries with their own forces or secret agencies such as the CIA and the KGB, they may do so indirectly through countries such as South Africa and Cuba.

Through a variety of financial, economic and military methods the superpowers and their allies in the First and Second Worlds have considerable impact on the destinies of Third World countries. Because of this they are accused of a new style of colonialism, given the name neo-colonialism. One of the aims of the non-aligned countries is to avoid the worst consequences of this dependency or interference.

1   From examples in this book, or recent reports in the media, give an account of superpower involvement in Latin America, Africa and Asia.
2   In what sense are all areas of the world effectively part of either the USA's or the Soviet Union's 'empire'?

Exercises on page 233
Military advisers and bases

Contras to get US arms next month

Nicaraguan rebels can expect to start receiving US military supplies after Senate approval of President Reagan's $100 million Contra aid request.

The priority will be training the Contras. The emphasis will be on special units and the officer corps. Much of the training will be conducted openly in Honduras by US Army special forces mobile training teams and Vietnam veterans on contract to the US Army.

President Reagan hailed the Senate action as a vote for democracy in Nicaragua.

Nicaragua's ambassador to the US said the Senate was giving its consent to 'a proxy war against the people and government of Nicaragua'.

The extract details US aid to Nicaraguan rebels. Nicaragua is one of the few Central American countries with a left-wing government. The photograph shows a boy working on a tobacco co-operative in Nicaragua

4.12 Summary

By the late 1980s Hong Kong was an oddity; a British colony but with a highly developed economy and a major trading nation in its own right

Exercises on page 234
Aspects of dependency in Southern Africa

☆ Most of the countries regarded as being part of the Third World were parts of European empires until quite recently. This shaped their economies and trade.

☆ The consequences of foreign control continue to be important. Some have been of benefit to the peoples of the former colonies and dependencies, but often they have proved harmful to well-being and continue to be a disadvantage to development.

☆ The cultural and political mix of many Third World countries is one of the consequences of artificial boundaries being drawn around territories by outside nations.

☆ Third World countries are part of global financial and economic systems, but due to their general lack of power and influence they tend to gain relatively little from their involvement.

☆ Because of their relative lack of economic development, or limited gains from development that has occurred, Third World countries lack capital for reasonable welfare programmes and economic growth. They have to borrow from wealthier nations and institutions.

The increasing problems of debt are partly explained by internal mismanagement, but more frequently result from sudden falls in income from the sale of commodities and rapid increases in interest rates for loan repayments.

☆ Development aid may be officially provided by governments either bilaterally or through international agencies. Smaller amounts are provided by voluntary non-governmental agencies. Motives for giving official aid are varied, and include the self-interest of the donors. Some aid is misused and misappropriated by officials and the more prosperous, and much fails to benefit those who need it most. There is disagreement about the relative emphasis that should be given to aid for welfare and aid for economic growth.

☆ Third World countries are seen as potential allies or threats by the more powerful capitalist and communist power blocs. The power blocs accordingly influence and interfere in Third World countries in a variety of ways. The non-aligned movement tries to resist such neo-colonialism, although some countries have marked sympathies with one or other superpower.

Chapter 5

Prospects for global development

To what extent are conditions of life in Third World countries essentially different from those in countries in other parts of the world? What prospects are there for development in Third World countries that will lead to higher standards of living and improved quality of life?

Water towers in Kuwait

5.1   Is there a Third World?

The dangers of labelling

Throughout this book there are examples of the great diversity of environments, standards of living and quality of life around the world. This is also shown daily in newspapers and on television. But only a tiny part of the whole world is known to most people, and there are huge gaps in our knowledge and understanding of the world and the way it works.

In an attempt to make sense of the many different environments and human ways of life we produce simplified pictures and explanations based on all the evidence available. In order to describe and explain contrasts in economic, social and political development, for example, some people have suggested that the nations of the world can be put into one of three groups. These have been called the First, Second and Third Worlds. The meanings of these terms were discussed on page 20, and many references have been made in this book to the Third World.

The First World is seen to consist largely of the countries of Europe, North America, Australia, New Zealand and Japan. In general these are industrialised, wealthy and powerful nations, and are run by democratically elected governments with capitalistic type economies. But it is obvious to everyone that the countries within, say, Europe are very different indeed, and to say that someone or some country is European has only limited value as a description.

Those countries with a centralised, state-run economy that are relatively industrialised, wealthy and powerful on a world scale are said to belong to the Second World. This group consists of the USSR and many of the countries of Eastern Europe.

The remaining countries are said to form the Third World. They have a variety of forms of government but tend to be poorer, only partially industrialised, and have large proportions of their total population experiencing great poverty. They also tend to be economically or politically dependent on other countries. This includes most countries of Latin America, Africa and Asia.

As with the First World, so with the Second and Third, there are great contrasts between the countries within them. There are equally great contrasts within the countries themselves. There is a strong tendency to think everything is like the average, or one or two limited examples. That produces inaccurate and stereotyped views of the world, and can lead to great misunderstanding and unjustified attitudes to other countries and their people.

Homeless people in New York, one of the world's richest cities, queue up for food

40 million Americans still go to bed hungry

Some nutrition experts estimate that perhaps 40 million Americans are hungry or inadequately fed.

The report 'Hunger in American Cities' confirmed that the number of destitute Americans is rising rapidly. A survey of eight cities found that the demand for emergency food and shelter is up between 100 and 500 per cent during the life of the Reagan Administration.

The report finds that those in soup lines are no longer temporary drop-outs of yesteryear, but also people who until recently had jobs and homes, and to whom accepting charity is almost as painful as starving.

Descent into poverty can be very rapid in the United States, and the chief sufferers are, as ever, the minorities, with 35.6 per cent of blacks calculated to be below the poverty line, mainly because of a vast increase in fatherless families.

Perhaps the most frightening statistic is that there are 13.5 million children living in poverty.

Differences and similarities

The few illustrations and extracts on these pages show the differences that exist within a First, Second and Third World country. The USA is unquestionably one of the richest and most powerful countries in the world, yet within it there is much poverty and unemployment. The USSR is likewise one of the most powerful and economically developed countries in the world. It prides itself on its caring attitude to people and the right of its people to work. Yet there is a control of thinking, expression and behaviour that many people in First and Third World countries would regard as a barrier to real quality of life.

Having recognised that differences exist, there is no question that some countries are similar in that they have average levels of wealth far below the world average. In these countries large proportions of the population are severely deprived in terms of nutrition, health, housing, employment and education. Such countries often seem unable to make any significant improvements to the lives of their peoples because of difficulties such as harsh environments, economic or political interference and even control by other countries, internal strife, or lack of capital or expertise to take the first steps. These countries can be grouped together because of common characteristics, and they are the countries referred to as the Third World.

Some statistical indicators of development for three of the world's largest and most populous countries

Country	GNP per capita (US $) 1985	Pecentage of adults literate 1985 male/female	Life expectancy at birth (years) 1985	Infant mortality rate (0–1 year) per thousand 1985
USA	16 690	99/99	75	11
USSR	4 550	99/98	72	24
India	270	57/29	57	105

The USSR is a country with many ethnic minorities. This photograph shows a demonstration in Armenia over a territorial dispute with its neighbouring Soviet Republic of Azerbaijan. This sort of demonstration is probably rare in the USSR

It is worth remembering that beneath these differences of wealth and living standards, and differences of belief and behaviour, all people share the same features of being human – they are born, grow up and die. One of the strange features of the world is how people can be exactly the same as human beings yet at the same time hold remarkably different beliefs and experience dramatically different ways of life and standards of living.

1   Use the statistics, photographs and extract to comment on 'The dangers of generalising, and of describing places and the conditions of people from limited evidence'.
2   Obtain or suggest evidence about life in Britain in the late 1980s that illustrates how difficult it is to make generalisations about a country.
3   Describe all the similarities you can think of between the life of a person in Britain and Bangladesh.

The entrance to the Taj Mahal Hotel in Bombay, India. There is wealth in the Third World

5.2 Alternative styles of development

Not all people have the same beliefs about what is desirable progress, and so it is not surprising that there are differing views about the aims of development. It is widely believed by people in countries of the First World that continued economic growth is the prime goal. The aim is to produce more from available resources using efficient and new technologies. There is considerable argument about whether the motive for this increased wealth is mainly to provide profits for those who invest money in wealth creation, or whether it is to provide a higher standard of living for all. There is no reason why both aims should not be the reason for greater effort to produce more.

'Appropriate development'

It is usually assumed by people in the so-called developed world that the Third World countries should try to copy them and create economies based on the investment of masses of capital and the building of large-scale industries. Their belief is that if people in the Third World were properly trained, worked hard enough and avoided waste and inefficiency, then in time they would become as prosperous as the First World. Some Third World governments and industrialists have been persuaded of this argument, and have devoted large amounts of development aid and foreign investment to large-scale and capital-intensive power and manufacturing industry. The wealth created, it is claimed, will 'trickle down' to the poor and needy. There are two important doubts. The first is whether that style of economic growth is what will really work in Third World countries. The second is whether, even if it did succeed, it would lead to a better quality of life in the future.

Third World countries lack the sort of capital that is needed for any massive project such as building a hydro-electric power station or a steelworks. So it either has to be asked for as aid, borrowed from commercial banks or received as investment from foreign companies that would hope to make a profit from the development. Whichever method of raising money is used, but especially the last two, the borrowing country comes under the economic influence or control of a foreign company or country. Very often, as has been shown, vast sums of interest have to be paid as well as the repayment of the

Advertisements for consumer goods in China – until recently a very rare sight. Communist China has followed its own style of development

Cooking by 'biogas' in Sri Lanka; a good example of appropriate technology

economic growth is a dangerous goal in all parts of the world. There are limited resources in the world, and although there are many that remain undeveloped, and unknown uses of many others yet to be discovered, sooner or later these will be used up if present practices continue. This is why there has been an increasing emphasis on avoiding wasteful use and practices, on recycling, and on conservation of resources. This is of little interest to those who are only concerned with immediate needs, but it is of great concern to those who want to maintain healthy environments and who feel a responsibility towards future generations. They argue that continued economic growth is not acceptable if it means the ultimate destruction of habitats and environments and the absolute exhaustion of the world's resources on which human life depends.

original capital borrowed. Even if the project succeeds, and the power station or steelworks or factory is built, there is no certainty that the majority of people will benefit. By its very nature modern large-scale industry may not be a big employer of labour, so only a relatively few new jobs may be created. Because there is usually so little spending power within the country the product may be manufactured mainly for foreign markets, with only limited internal sales to a few wealthier farmers or families.

It has been strongly argued for decades now that there is no standard method of development that can be applied world-wide. Large-scale industrial development using modern technology may be the least likely form of development to lead to general improvements in living standards and quality of life. Smaller-scale developments that depend on simpler technology and lead to far higher employment may well generate much greater benefits for the majority of a community or a country. This approach of adjusting new activities to the special circumstances of particular places and people is frequently known as 'appropriate development'.

'Sustainable development'

Not only is it argued that development activity needs to be adjusted to the needs of particular Third World areas, but also that the very idea of never-ending

1 What are the arguments for a) bigger and bigger mines, factories, power stations, offices, businesses and organisations, and for b) small-scale methods of production and organisation of economic, social and political life?
2 Say whether you think the warnings of 'limits to growth' are scaremongering, or a sign of a real resource and environmental crisis.

Billboard and squatter's hut in Bombay, India. The poverty of the hut contrasts sharply with the glamour portrayed by the film poster

5.3 Interdependence and the Third World

This crowded breakfast table contains many foods and ingredients from the Third World. Everytime we sit down to eat we should be reminded of our dependence on Third World products

This home in the Philippines, with its mixture of local and Western styles, provides an illustration of cultural interdependence. On the television is an American show

Interconnections

In this book, and every day in news-papers and on television, there are examples of the interdependence of places and peoples. One of the re-markable features about the world is the way in which everything is inter-connected. In any environment, for example, there is a close relationship between landforms, rocks, soils, veget-ation and animal life. Weather and climate also influence the type of habitat and the ecology of any place, while weather and climate are caused by the major movements of air around the surface of the earth. Over the centuries human behaviour has been strongly influenced by these environmental features, although more recently people have developed the skills and tech-nologies to dramatically transform environments.

The earth is a complicated system of related parts. This can be more easily imagined when it is seen from space, with the changing pattern of cloud cover and associated weather affecting the earth and life below. Over billions of years this system has achieved a sort of balance or equilibrium, so that a change in one part leads to related changes in many others. This is one of the dangers of massive interference in the environ-ment by human behaviour. The earth itself is part of the solar system which in turn is part of the universe, and the universe is only partly known and little understood. Seen from this point of view – and there are other very different views of the world – this earth and the life on it is only one part of a much vaster whole. And human life has only existed on it for a short period of time.

There are interconnections between people as well as those linking people and the environment. Probably the most basic are those between parents and children, members of families and close personal friends. Then there are the many relationships based on working and social activities. In some cases the relationship can be one of extreme dependency, such as that experienced by severely ill or handicapped people. Sometimes the relationship is of a very different kind, such as that between prisoners and their warders or jailers. Millions of these interconnections and dependencies affect the lives of people throughout the world every minute of the day.

Interdependence between nations

The sorts of relationships mentioned above also exist between groups of people and nations, and can be described as economic, social and political inter-dependence. Economic interdependence at this level is illustrated by the trade and exchange of goods that takes place between different parts of the world. In many cases both partners in a trading arrangement get some benefit from it, but both in the past and the present there are examples where one partner seems to gain by far the most. There is resentment in some former colonies, for example, that when they were con-trolled by other more powerful nations their raw materials and resources were taken for very little in return. Nowadays

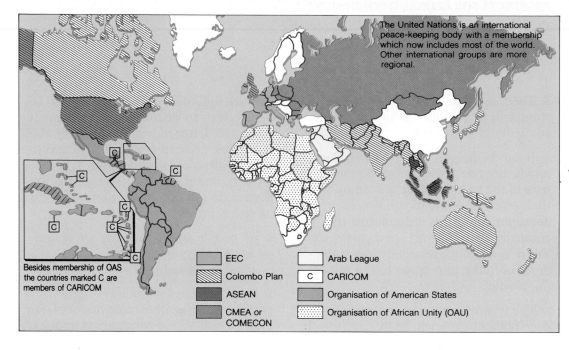

The United Nations is an international peace-keeping body with a membership which now includes most of the world. Other international groups are more regional.

	EEC		Arab League
	Colombo Plan	C	CARICOM
	ASEAN		Organisation of American States
	CMEA or COMECON		Organisation of African Unity (OAU)

Besides membership of OAS the countries marked C are members of CARICOM

Major world political and economic groupings. Other maps in this book show NATO, the Warsaw Pact countries, the Commonwealth, ECOWAS, and OPEC

inequalities of trade are more bound up with international finance, wild changes in commodity prices and heavy debt repayments. Many countries co-operate closely in economic matters and even organise themselves into economic unions, such as the European Economic Community (EEC).

There are many instances of social interdependence. Some result from the large migrations of people that have taken place in recent decades, with families becoming spread around the world. Others result from travel and tourism, and the use of television and other forms of communication on a global scale. Social interdependence can also be one-sided, with some people and nations more able to move about the world and manage the flow of inform-ation and cultural influence than others.

Political interdependence often involves countries joining together for defence against possible or actual at-tackers. Most nations of the world belong to a political or military group to share in the provision of armed forces for defence. A very different form of political dependency is where one nation or group is controlled by another through the use of military force, and so lacks the opportunity to control its own affairs.

Interdependence, then, takes many forms. It can be between groups within a country, between nations or between groups of nations. The relationship can

be healthy and of equal benefit to all, or one-sided and of benefit largely to the most influential and powerful only. In extreme cases it can mean control and exploitation. There is much evidence to suggest that Third World countries gain least from this interdependence, and that the biggest hope for an improvement in their condition would be changed attitudes and a little more help from others in this interdependent world.

1 Give detailed examples of economic, social and political interdependence involving Britain and British people.
2 Give examples from this book where Third World countries have a) gained and b) suffered from economic, social and political interdependence.

FAST WITH OXFAM

On November 7, 8 and 9 we'll be Fasting For Change to raise urgently needed funds for our development work overseas. At the same time we will be saying to our government:
● It's time to end the debt crisis.
● It's time for fair trade.
● It's time for aid that **prevents** hunger.
● It's time for us to become part of the solution instead of part of the problem.
Please join us.
DON'T STOP THE GIVING – STOP THE TAKING

This Oxfam advertisement, from *The Observer,* 26 October 1986, stresses the interdependence between the developed world and the Third World

5.4 Summary

☆ There are great differences in living standards and quality of life within all countries of the world.

☆ It can be helpful, as long as all the pitfalls are recognised, to group countries of the world loosely according to their levels of economic and social development. If this is done then the countries usually considered part of the Third World have relatively low standards of living in relation to the rest of the world.

☆ The type of economic and social development experienced by countries in the First and Second Worlds may not be appropriate for Third World countries. Large-scale and capital-intensive manufacturing industries and huge power projects may not greatly benefit large sections of the population.

☆ There are many signs that the goal of unlimited economic growth may be harmful in the more developed First and Second Worlds. Environmental damage, resource exhaustion and habitat and wildlife destruction may have disastrous effects on the well-being of everyone in the longer term.

☆ More and more thought and effort is being given to sustainable development – the avoidance of waste, recycling of resources, efficient use of energy and the conservation of environments. Economic growth, and the distribution of wealth from such growth, needs to be related to its global impact.

☆ All people and all nations are bound up in global economic, social and political relationships.

☆ Interdependence can have both positive and negative consequences. For many Third World countries interdependence does not work in their favour.

☆ Prospects for global development depend on changed attitudes to economic growth and the distribution of wealth, the conservation of resources and environments, and on fairer rewards from interdependence.

Good research facilities are vital for Third World countries and yet are in very short supply

Part Two   The Third World: case studies in development

Atlas section

Case studies

The many factors leading to differences in economic and social development do not operate in isolation. They interact at particular places in various combinations and to varying levels. Development is also strongly influenced by decisions made by governments. These case studies attempt to describe how the processes of development were operating in eight Third World countries in the late 1980s. They provide illustrations of the ideas explained in Part One.

People and places dealt with in the case studies: Ethiopia (left), Bolivia (above left) and Saudi Arabia (above)

Political

• Capital city

— International boundary

*after a country name indicates that
country and capital have the same name

Modified Gall Projection

Equatorial Scale 1 : 88 000 000

Abbreviations

A.	ANDORRA*
ALB.	ALBANIA
AUST.	AUSTRIA
BELG.	BELGIUM
CENT. AF. REP.	CENTRAL AFRICAN REPUBLIC
CZECH.	CZECHOSLOVAKIA
EQ. GUINEA	EQUATORIAL GUINEA
F.R.G.	FEDERAL REPUBLIC OF GERMANY
G.D.R.	GERMAN DEMOCRATIC REPUBLIC
L.	LIECHTENSTEIN
LUX.	LUXEMBOURG *
M.	MONACO
NETH.	NETHERLANDS
SWITZ.	SWITZERLAND
U.A.E.	UNITED ARAB EMIRATES
U.S.A.	UNITED STATES OF AMERICA
U.S.S.R.	UNION OF SOVIET SOCIALIST REPUBLICS
YEMEN P.D.R.	YEMEN PEOPLES DEMOCRATIC REPUBLIC

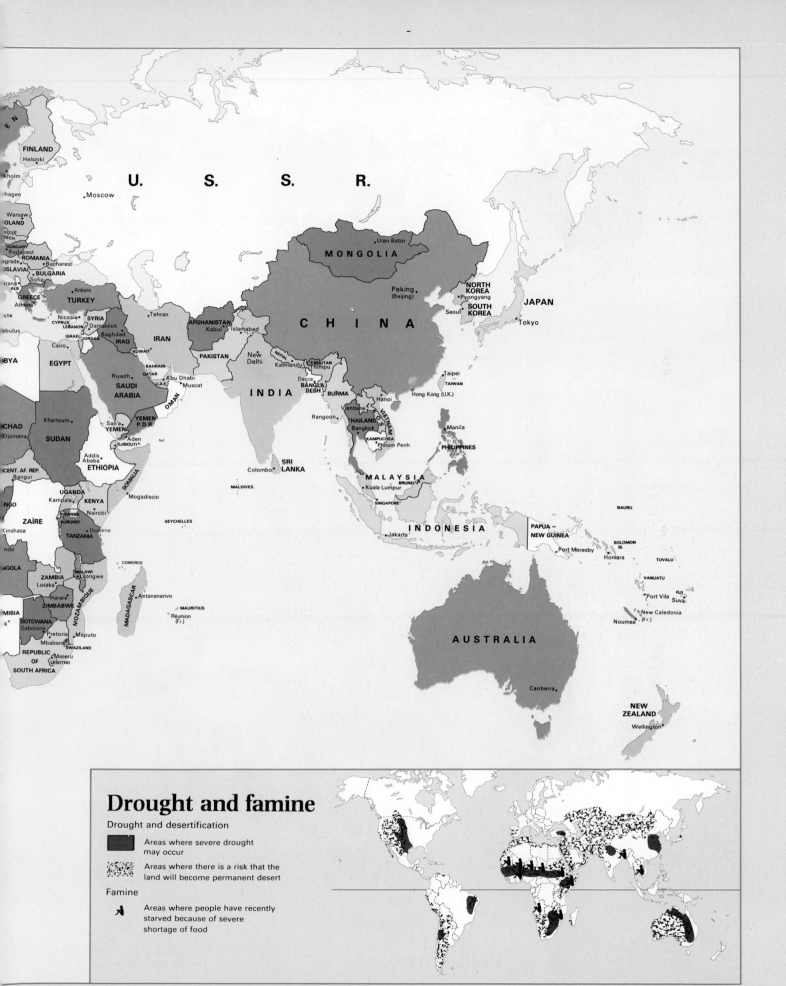

Drought and famine

Drought and desertification

Areas where severe drought may occur

Areas where there is a risk that the land will become permanent desert

Famine

Areas where people have recently starved because of severe shortage of food

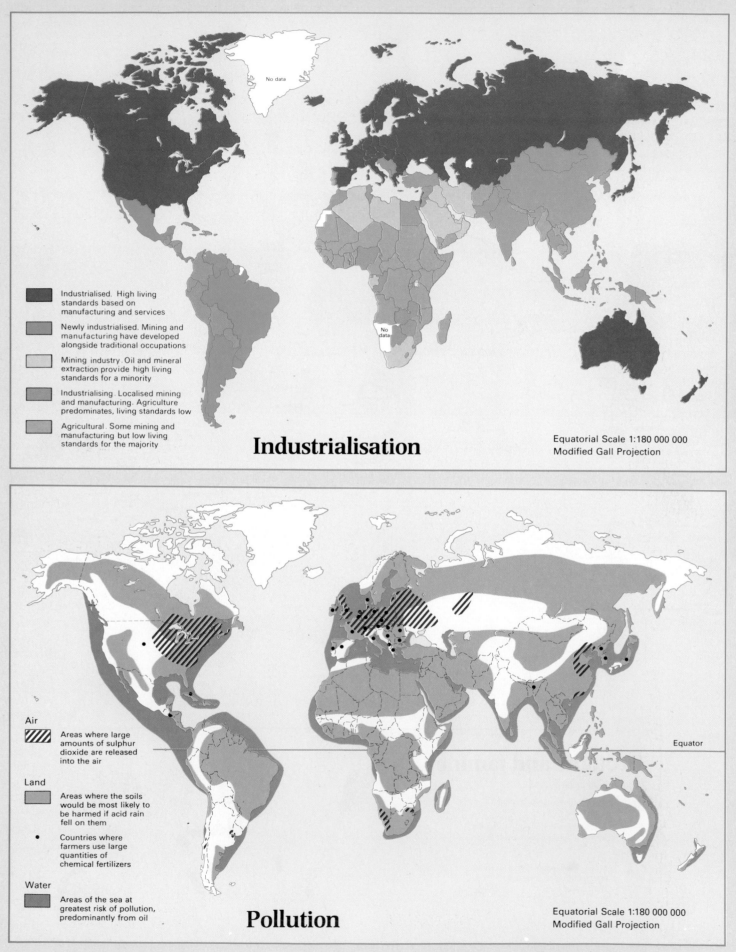

Industrialisation

Industrialised. High living standards based on manufacturing and services

Newly industrialised. Mining and manufacturing have developed alongside traditional occupations

Mining industry. Oil and mineral extraction provide high living standards for a minority

Industrialising. Localised mining and manufacturing. Agriculture predominates, living standards low

Agricultural. Some mining and manufacturing but low living standards for the majority

No data

Equatorial Scale 1:180 000 000
Modified Gall Projection

Pollution

Air

Areas where large amounts of sulphur dioxide are released into the air

Land

Areas where the soils would be most likely to be harmed if acid rain fell on them

Countries where farmers use large quantities of chemical fertilizers

Water

Areas of the sea at greatest risk of pollution, predominantly from oil

Equator

Equatorial Scale 1:180 000 000
Modified Gall Projection

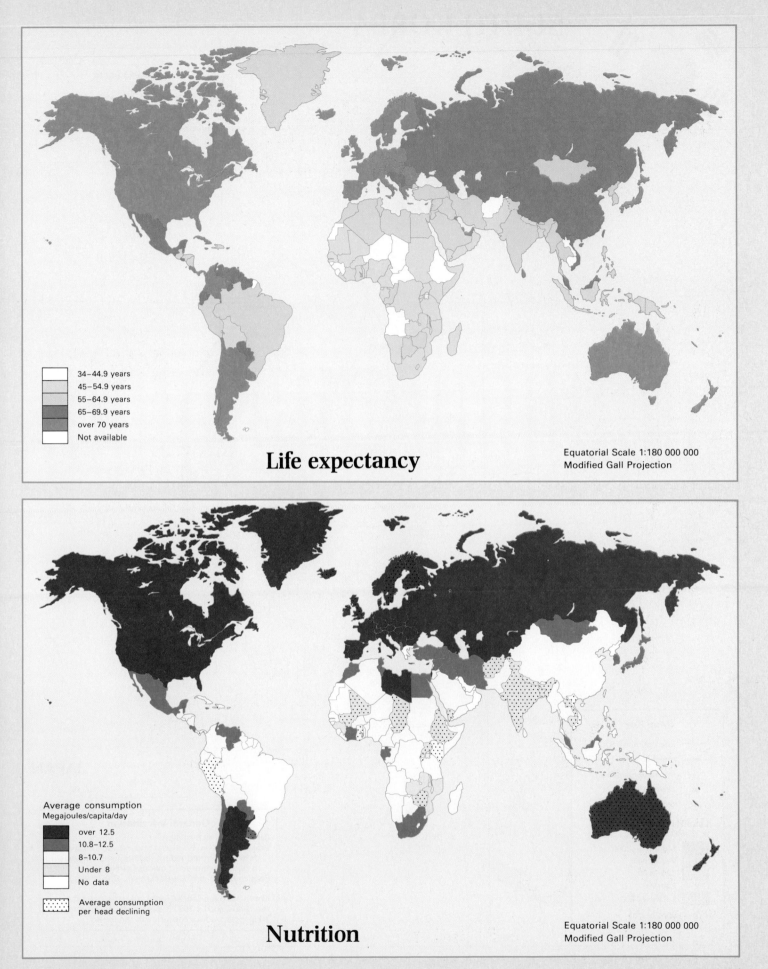

Life expectancy

34–44.9 years
45–54.9 years
55–64.9 years
65–69.9 years
over 70 years
Not available

Equatorial Scale 1:180 000 000
Modified Gall Projection

Nutrition

Average consumption
Megajoules/capita/day

over 12.5
10.8–12.5
8–10.7
Under 8
No data

Average consumption
per head declining

Equatorial Scale 1:180 000 000
Modified Gall Projection

SOUTH KOREA

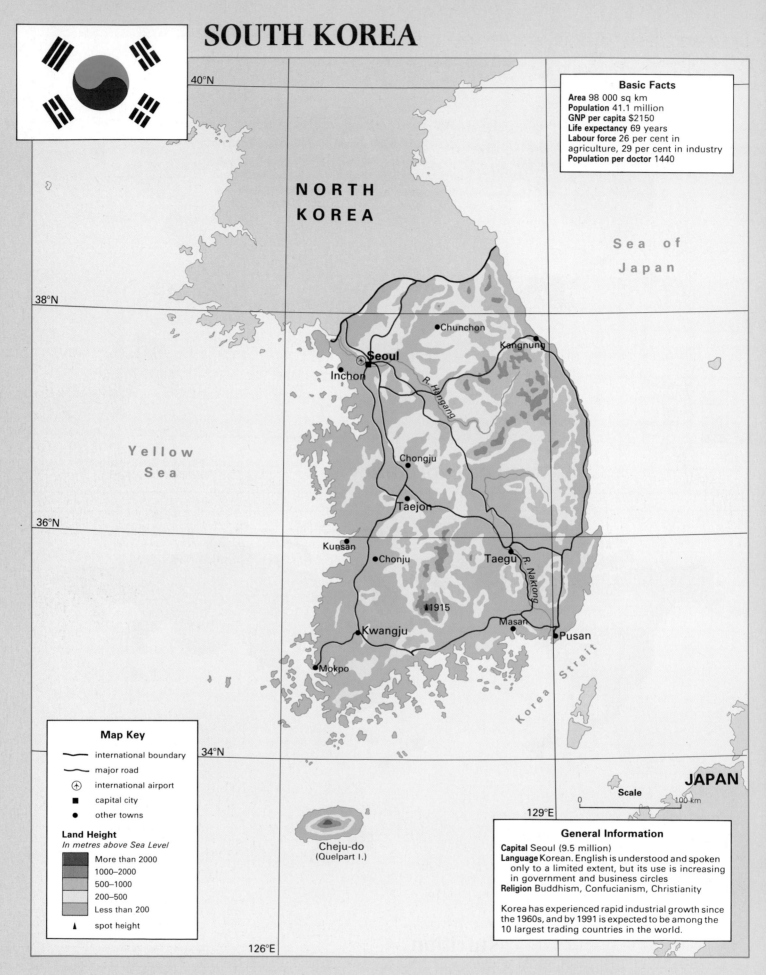

Basic Facts
Area 98 000 sq km
Population 41.1 million
GNP per capita $2150
Life expectancy 69 years
Labour force 26 per cent in agriculture, 29 per cent in industry
Population per doctor 1440

NORTH KOREA

Sea of Japan

Chunchon

Kangnung

Seoul

Inchon

R. Hangang

Yellow Sea

Chongju

Taejon

Kunsan

Chonju

Taegu

R. Naktong

▲1915

Masan

Kwangju

Pusan

Korea Strait

Mokpo

Cheju-do (Quelpart I.)

JAPAN

Map Key
— international boundary
— major road
⊕ international airport
■ capital city
● other towns

Land Height
In metres above Sea Level
More than 2000
1000–2000
500–1000
200–500
Less than 200
▲ spot height

Scale
0   100 km

General Information
Capital Seoul (9.5 million)
Language Korean. English is understood and spoken only to a limited extent, but its use is increasing in government and business circles
Religion Buddhism, Confucianism, Christianity

Korea has experienced rapid industrial growth since the 1960s, and by 1991 is expected to be among the 10 largest trading countries in the world.

40°N
38°N
36°N
34°N
126°E
129°E

INDONESIA

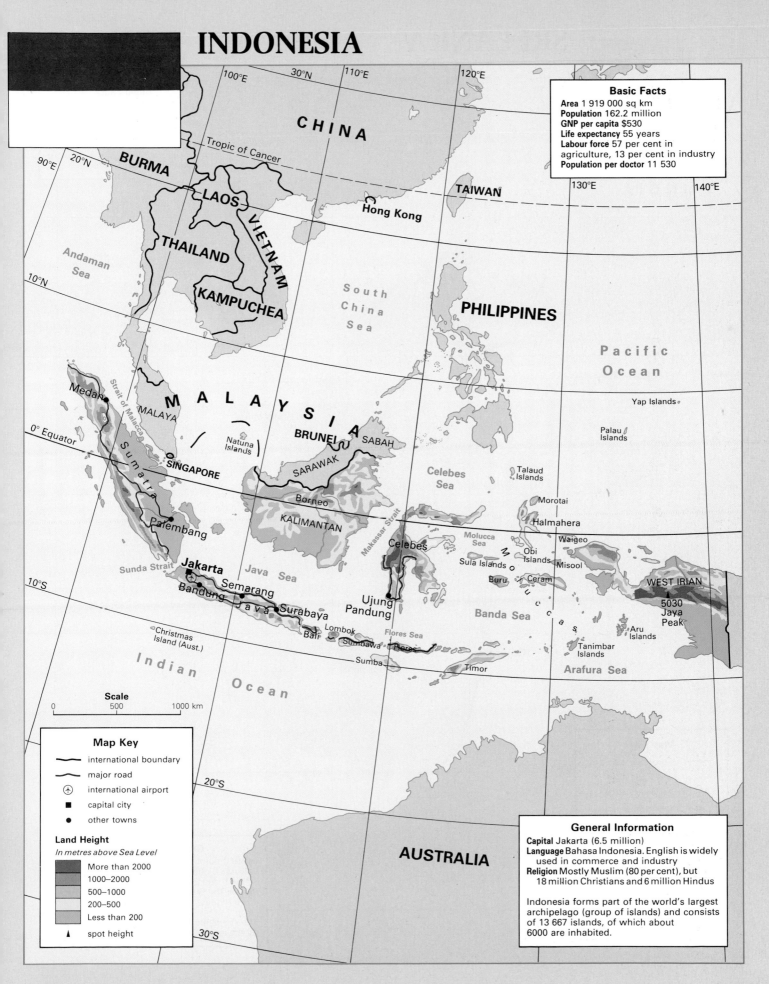

Basic Facts
Area 1 919 000 sq km
Population 162.2 million
GNP per capita $530
Life expectancy 55 years
Labour force 57 per cent in agriculture, 13 per cent in industry
Population per doctor 11 530

CHINA

Tropic of Cancer

BURMA

LAOS

VIETNAM

THAILAND

KAMPUCHEA

Andaman Sea

South China Sea

PHILIPPINES

TAIWAN

Hong Kong

Pacific Ocean

Yap Islands

Palau Islands

MALAYSIA

MALAYA

Medan

Strait of Malacca

Natuna Islands

BRUNEI   SABAH

SARAWAK

Borneo

KALIMANTAN

Celebes Sea

Talaud Islands

Morotai

Halmahera

Waigeo

Molucca Sea

Obi Islands

Misool

Sula Islands

Ceram

Buru

SINGAPORE

Equator   0°

Sumatra

Palembang

Makassar Strait

Celebes

Ujung Pandung

Banda Sea

WEST IRIAN

5030 Jaya Peak

Aru Islands

Sunda Strait

Jakarta

Bandung   Semarang

Java   Surabaya

Bali   Lombok

Sumbawa   Flores

Sumba

Java Sea

Flores Sea

Christmas Island (Aust.)

Timor

Tanimbar Islands

Arafura Sea

Indian   Ocean

Scale
0   500   1000 km

Map Key
- ⌒ international boundary
- ⌒ major road
- ⊕ international airport
- ■ capital city
- ● other towns

Land Height
In metres above Sea Level
- More than 2000
- 1000–2000
- 500–1000
- 200–500
- Less than 200
- ▲ spot height

AUSTRALIA

General Information
Capital Jakarta (6.5 million)
Language Bahasa Indonesia. English is widely used in commerce and industry
Religion Mostly Muslim (80 per cent), but 18 million Christians and 6 million Hindus

Indonesia forms part of the world's largest archipelago (group of islands) and consists of 13 667 islands, of which about 6000 are inhabited.

117

SRI LANKA

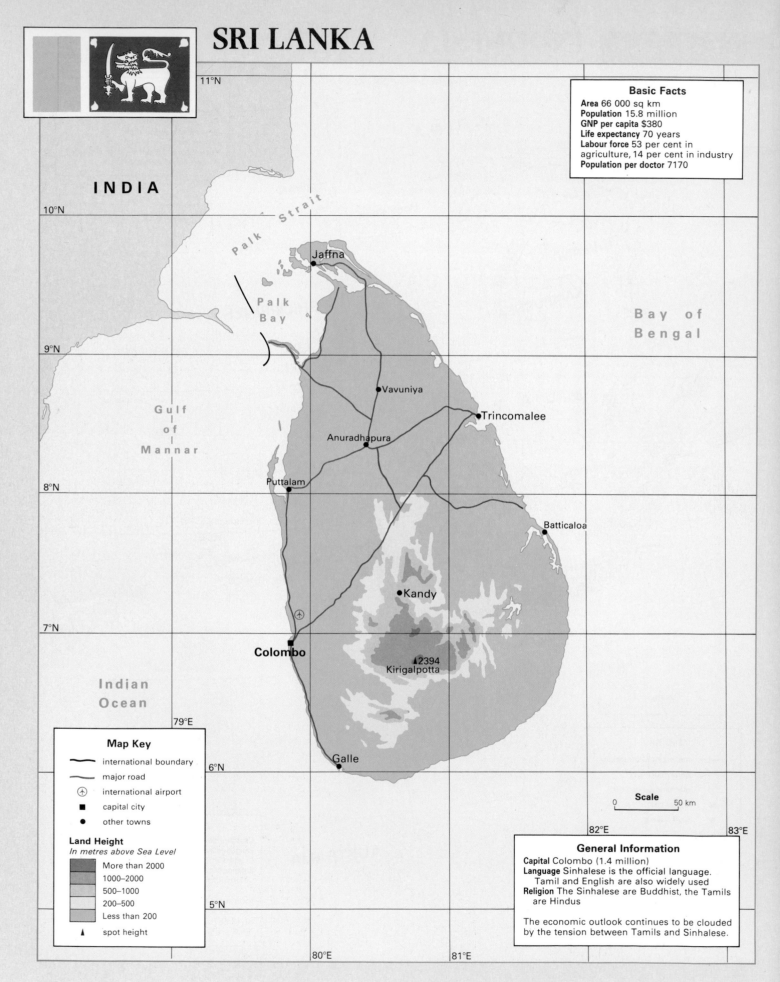

INDIA

Basic Facts
Area 66 000 sq km
Population 15.8 million
GNP per capita $380
Life expectancy 70 years
Labour force 53 per cent in agriculture, 14 per cent in industry
Population per doctor 7170

11°N

10°N

Palk Strait

Jaffna

Palk Bay

9°N

Bay of Bengal

Vavuniya

Gulf of Mannar

Trincomalee

Anuradhapura

8°N

Puttalam

Batticaloa

Indian Ocean

Kandy

7°N

Colombo

▲2394
Kirigalpotta

79°E

6°N

Galle

Map Key

⸺ international boundary
⸺ major road
⊕ international airport
■ capital city
● other towns

Land Height
In metres above Sea Level
More than 2000
1000–2000
500–1000
200–500
Less than 200
▲ spot height

Scale
0 ———— 50 km

82°E   83°E

General Information
Capital Colombo (1.4 million)
Language Sinhalese is the official language. Tamil and English are also widely used
Religion The Sinhalese are Buddhist, the Tamils are Hindus

The economic outlook continues to be clouded by the tension between Tamils and Sinhalese.

5°N

80°E   81°E

SAUDI ARABIA

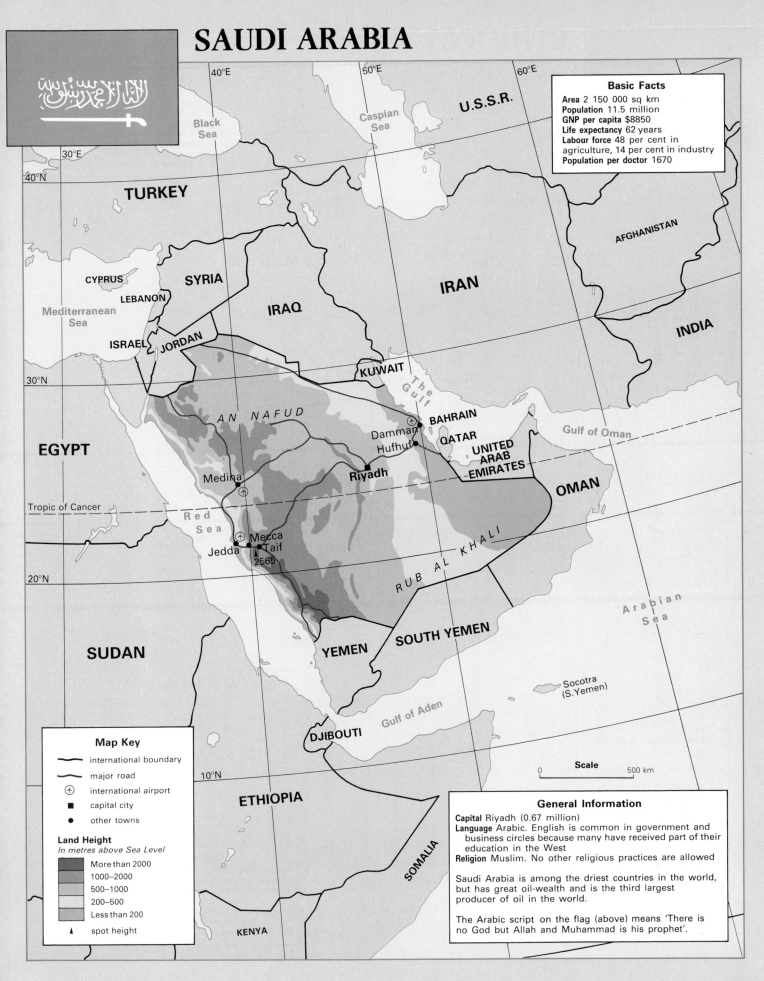

Basic Facts

Area 2 150 000 sq km
Population 11.5 million
GNP per capita $8850
Life expectancy 62 years
Labour force 48 per cent in agriculture, 14 per cent in industry
Population per doctor 1670

TURKEY

Black Sea

Caspian Sea

U.S.S.R.

CYPRUS

SYRIA

LEBANON

IRAQ

IRAN

AFGHANISTAN

Mediterranean Sea

ISRAEL

JORDAN

KUWAIT

INDIA

EGYPT

AN NAFUD

The Gulf

BAHRAIN

Damman
Hufhuf

QATAR

Gulf of Oman

UNITED ARAB EMIRATES

OMAN

Medina

Riyadh

Tropic of Cancer

Red Sea

Mecca
Jedda   Taif
2565

RUB AL KHALI

Arabian Sea

SUDAN

YEMEN

SOUTH YEMEN

Socotra (S.Yemen)

DJIBOUTI

Gulf of Aden

ETHIOPIA

SOMALIA

KENYA

Scale

0   500 km

Map Key

— international boundary

— major road

⊕ international airport

■ capital city

● other towns

Land Height

In metres above Sea Level

More than 2000

1000–2000

500–1000

200–500

Less than 200

▲ spot height

General Information

Capital Riyadh (0.67 million)
Language Arabic. English is common in government and business circles because many have received part of their education in the West
Religion Muslim. No other religious practices are allowed

Saudi Arabia is among the driest countries in the world, but has great oil-wealth and is the third largest producer of oil in the world.

The Arabic script on the flag (above) means 'There is no God but Allah and Muhammad is his prophet'.

ETHIOPIA

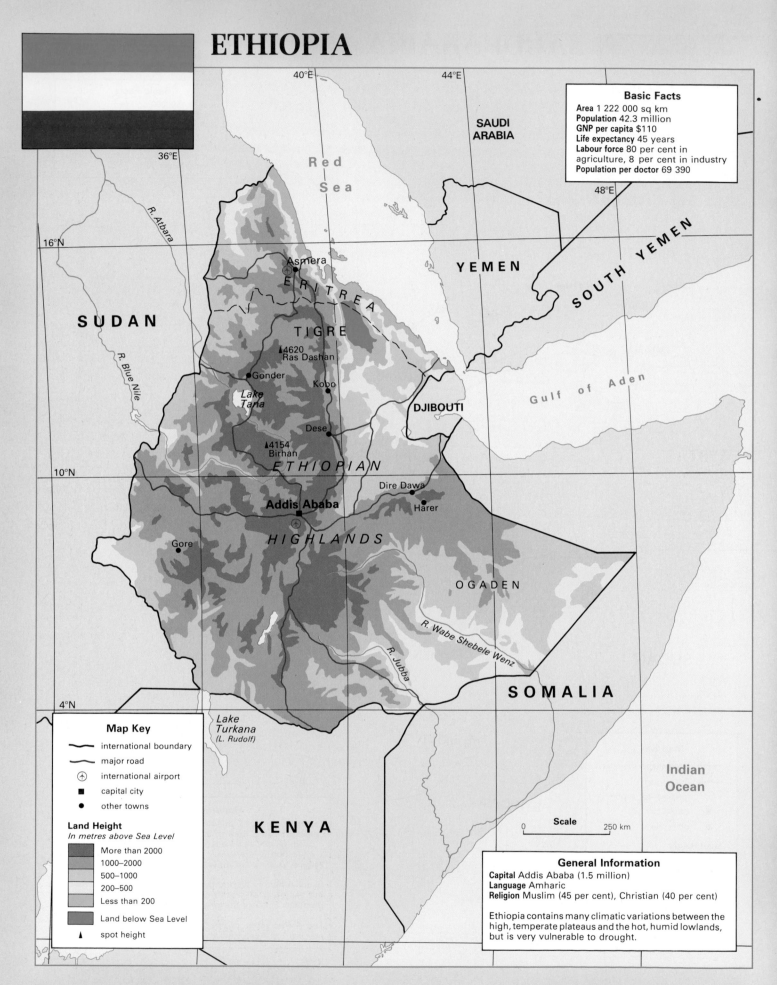

SAUDI ARABIA

YEMEN

SOUTH YEMEN

Red Sea

Gulf of Aden

SUDAN

R. Atbara

R. Blue Nile

⊕ Asmera

E R I T R E A

T I G R E

▲4620 Ras Dashan

● Gonder

● Kobo

Lake Tana

● Dese

▲4154 Birhan

E T H I O P I A N

● Dire Dawa

■ **Addis Ababa** ⊕

● Harer

H I G H L A N D S

● Gore

DJIBOUTI

O G A D E N

R. Wabe Shebele Wenz

R. Jubba

S O M A L I A

Lake Turkana (L. Rudolf)

K E N Y A

Indian Ocean

Basic Facts
Area 1 222 000 sq km
Population 42.3 million
GNP per capita $110
Life expectancy 45 years
Labour force 80 per cent in agriculture, 8 per cent in industry
Population per doctor 69 390

Map Key
⎯⎯ international boundary
⎯ major road
⊕ international airport
■ capital city
● other towns

Land Height
In metres above Sea Level
More than 2000
1000–2000
500–1000
200–500
Less than 200
Land below Sea Level
▲ spot height

Scale
0 ⎯⎯⎯⎯ 250 km

General Information
Capital Addis Ababa (1.5 million)
Language Amharic
Religion Muslim (45 per cent), Christian (40 per cent)

Ethiopia contains many climatic variations between the high, temperate plateaus and the hot, humid lowlands, but is very vulnerable to drought.

NIGERIA

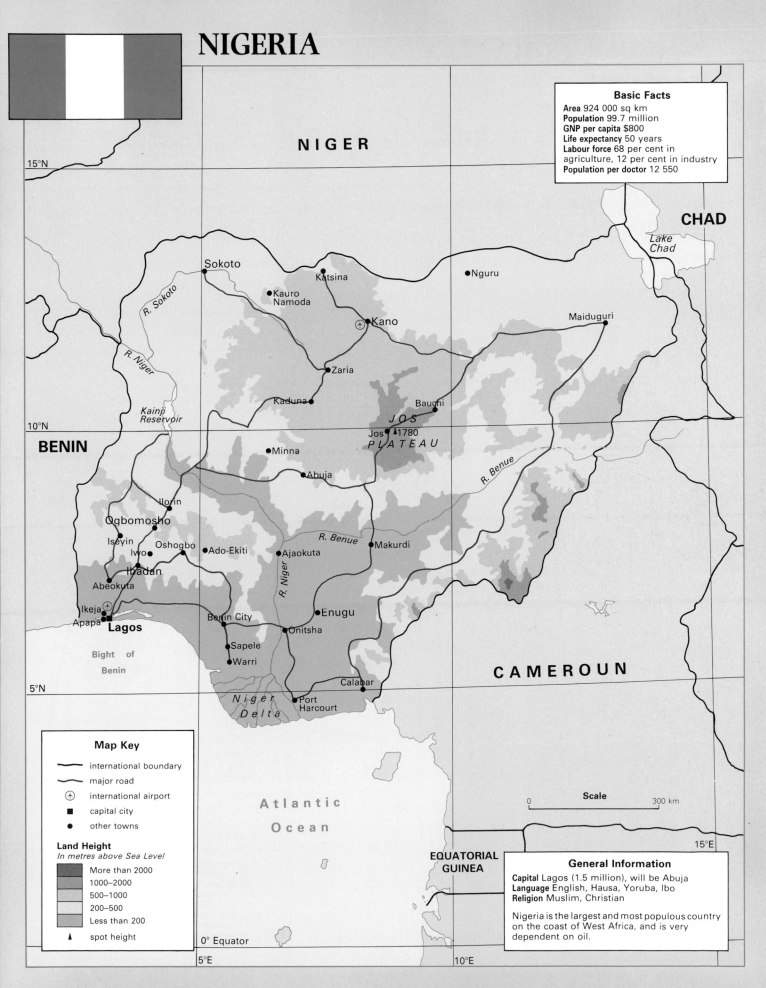

NIGER

CHAD

Basic Facts
Area 924 000 sq km
Population 99.7 million
GNP per capita $800
Life expectancy 50 years
Labour force 68 per cent in agriculture, 12 per cent in industry
Population per doctor 12 550

Lake Chad

15°N

Sokoto
R. Sokoto
•Katsina
•Nguru
Kauro Namoda
•Maiduguri
⊕Kano
R. Niger
Zaria
Kainji Reservoir
Kaduna•
Baughi•
10°N
JOS
Jos• ▲1780
PLATEAU
BENIN
•Minna
R. Benue
•Abuja
Ilorin•
Ogbomosho
Iseyin•
Oshogbo•
Iwo•
•Ado-Ekiti
R. Benue
Makurdi•
Ibadan•
Ajaokuta•
Abeokuta•
R. Niger
Ikeja⊕
Apapa•
Lagos ■
Benin City•
•Enugu
Onitsha•
Bight of Benin
Sapele•
•Warri
Calabar•
CAMEROUN
5°N
Niger Delta
Port Harcourt•

Atlantic Ocean

Map Key
— international boundary
— major road
⊕ international airport
■ capital city
• other towns

Land Height
In metres above Sea Level
More than 2000
1000–2000
500–1000
200–500
Less than 200
▲ spot height

EQUATORIAL GUINEA

15°E

General Information
Capital Lagos (1.5 million), will be Abuja
Language English, Hausa, Yoruba, Ibo
Religion Muslim, Christian

Nigeria is the largest and most populous country on the coast of West Africa, and is very dependent on oil.

0° Equator

5°E   10°E

Scale
0   300 km

CUBA

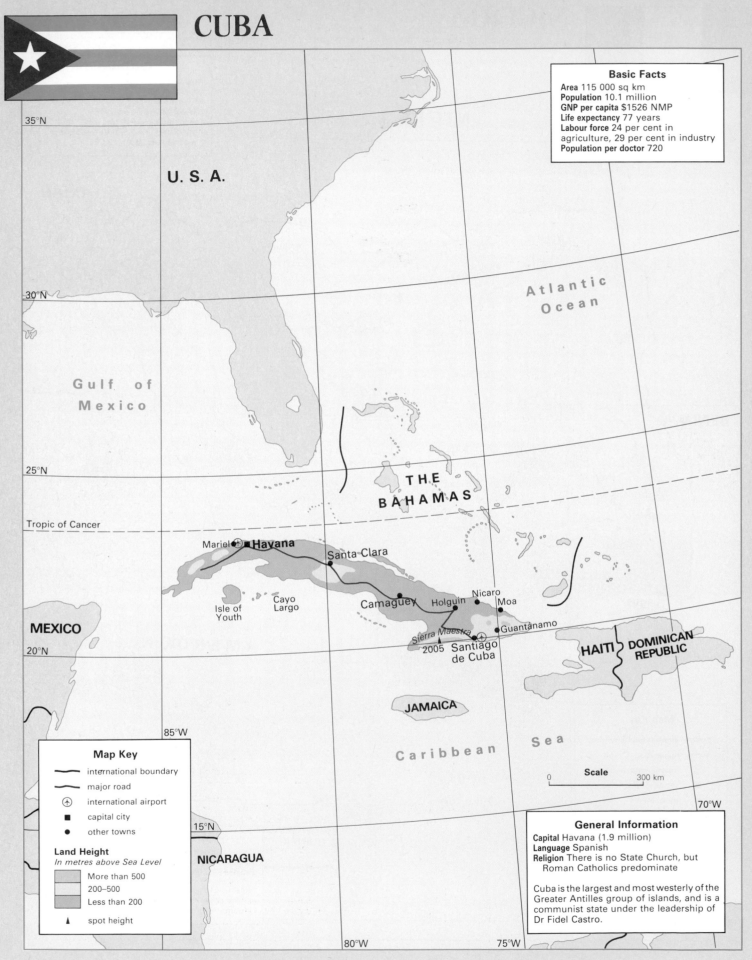

Basic Facts
Area 115 000 sq km
Population 10.1 million
GNP per capita $1526 NMP
Life expectancy 77 years
Labour force 24 per cent in
agriculture, 29 per cent in industry
Population per doctor 720

35°N

U.S.A.

30°N

Atlantic
Ocean

Gulf of
Mexico

25°N

THE
BAHAMAS

Tropic of Cancer

Mariel ⊕ ■ **Havana**

Santa Clara

Cayo
Largo

Isle of
Youth

Camaguey   Holguin   Nicaro   Moa

Sierra Maestra   Guantánamo

2005   Santiago
de Cuba

MEXICO

20°N

HAITI   DOMINICAN
REPUBLIC

JAMAICA

85°W

Caribbean   Sea

Scale
0   300 km

70°W

Map Key
⎯⎯ international boundary
⎯⎯ major road
⊕ international airport
■ capital city
● other towns

Land Height
In metres above Sea Level
More than 500
200–500
Less than 200
▲ spot height

15°N

NICARAGUA

80°W   75°W

General Information
Capital Havana (1.9 million)
Language Spanish
Religion There is no State Church, but
    Roman Catholics predominate

Cuba is the largest and most westerly of the
Greater Antilles group of islands, and is a
communist state under the leadership of
Dr Fidel Castro.

BOLIVIA

Basic Facts
Area 1 099 000 sq km
Population 6.4 million
GNP per capita $470
Life expectancy 53 years
Labour force 46 per cent in
   agriculture, 20 per cent in industry
Population per doctor 2100

12°S

B R A Z I L

PERU

Rio Madre de Dios

Rio Beni

Rio Guaporé

R. Mamoré

● Trinidad

Lake
Titicaca

16°S

▲6485

■ **La Paz**

A N D E S

Yungas

Genl.
Saavedra ●
Montero ●
Santa
Cruz ●

E A S T E R N

Altiplano

● Cochabamba

● Oruro

L O W L A N D S

CORDILLERA OCCIDENTAL

*Lago de
Poopó*

A N D E S

● Sucre

20°S

● Potosí

CHILE

● Tarija

P A R A G U A Y

Scale
0   200 km

24°S

A R G E N T I N A

Map Key

⎯⎯⎯ international boundary

⎯⎯⎯ major road

⊕ international airport

■ capital city

● other towns

Land Height
In metres above Sea Level

More than 5000
2000–5000
1000–2000
500–1000
200–500
Less than 200

▲ spot height

General Information

Capital La Paz (0.88 million)
Language Spanish is the language of the educated classes,
   Aymará or Quechua the language of the majority of Indians
Religion Roman Catholicism is the recognised religion of
   the state

Bolivia is a landlocked state (it had a coast and ports
until defeated by Chile in 1884).

Bolivia derives its name from its liberator, Simon
Bolivar (1783–1830).

For large-scale map, see page 116

A Korean farming family. The main crops are rice, barley, wheat, soyabeans and potatoes, grown mostly on smallholdings like this

The Republic of South Korea occupies the southern half of the Korean peninsula. China lies about 350 kilometres to the west across the Yellow Sea, and Japan some 150 kilometres to the south-east across the Sea of Japan. It is separated from North Korea by a de-militarised 'neutral' zone about four kilometres wide running approximately along the 38th parallel of latitude. With an area of 98 484 square kilometres it is a little bigger than Scotland, but its population of well over 42 million is about eight times as large.

Physical environment

A fairly large proportion of the country consists of high and rugged mountains. These run down the eastern side of the country, and there is only a narrow coastal plain and few settlements along the east coast. Lower ridges and valleys spread from the mountains towards the lowlands and wider coastal plains of the west and south-west. This coastline is deeply indented, providing sheltered harbours. Offshore there are many islands including the large volcanic one of Cheju-do which features the highest peak in Korea (it is far higher than Ben Nevis). In the winters the mountains of South Korea are covered with snow

brought by the cold north-westerly winds from continental Asia. In summer the weather is hot and humid, and the country suffers the same heavy monsoon rains and tropical storms as nearby Japan.

The people and their background

For most of their history North and South Korea were a single country. The original inhabitants are thought to have migrated from the continental interior via Siberia and Manchuria. The Koreans have a distinct culture and they are ethnically and culturally different from their Chinese and Japanese neighbours. For centuries the Koreans were ruled by kings from their capital city of Seoul. They had little contact with Western traders, missionaries or diplomats, partly because of the remoteness of the country and partly because they wished to remain isolated. However, both China and Japan were interested in gaining influence and power in the territory.

In 1895 Japanese forces fought and drove Chinese invaders out of Korea, and then the defeat of Russia in 1905 confirmed Japan's dominant position in the country. The last king was deposed in 1910 and Japan formally annexed and took control of the country. For the next 35 years Korea was a Japanese colony. It experienced the benefits and losses of all colonial countries, and there is still today much resentment towards the Japanese because of this past relationship.

Following the defeat of Japan in 1945 at the end of the Second World War, Korea was occupied by Soviet troops from the north and Americans from the south, the armies meeting roughly midway through the peninsula. Under the influence of the USSR the northern Koreans adopted a communist system of government, while the southern Koreans under the influence of the USA adopted a western style of democracy. A joint Soviet-American commission was asked to arrange elections

A night-time view of Seoul, a city of 10 million people (London has a population of 6.5 million). Seoul was host city for the 1988 Olympic Games

for the whole country, but when they failed the United Nations set up another commission to try and unify the country under one government. The North Koreans refused to take part in the elections, so the United Nations declared the government elected by the South Koreans to be the national government.

Understandably, the North Koreans would not accept this, and in 1950 they invaded South Korea. They were opposed by a United Nations force that drove them back towards the Chinese frontier. A large Chinese army then joined the North Koreans and began to drive the United Nations forces back into the south. An armistice was declared in 1953, and a demilitarised zone established roughly along the 38th parallel of latitude. This still exists, and marks the frontier between North and South Korea. There are about 45 000 American troops in South Korea, and one of their main tasks is to patrol the northern frontier to prevent a feared invasion from the north and take-over by a communist government.

The beginnings of economic development

After the Korean War massive economic and military aid was given by the Western nations, especially the USA. This took the form of cheap loans, technical assistance and military equipment. The first steps in economic development were based on this external support, but from about the mid-1960s development gathered pace due to the determined efforts and skills of the Korean people themselves. By successfully competing in the world economy South Korea has experienced phenomenal growth, making it a good example of a 'newly-industrialising' Third World country.

1   Explain the environmental and historical disadvantages that made economic development difficult in South Korea.
2   In what ways did the existence of a hostile communist neighbour to the north help in the economic development of South Korea?

South Korean children on a visit to the sixth century temple of Pulguk-sa

2 A 'miracle' of economic development

In spite of only a quarter of its area being suitable for farming, a lack of industrial raw materials, and a need to import almost all its energy in the form of oil, South Korea experienced remarkable economic growth after 1960. The statistics show that in 1984 the number of people employed in agriculture on the one hand and mining and manufacturing on the other was roughly the same (26 and 23 per cent). The previous years had seen a big drop in the numbers in farming, however, while those in mining and manufacturing had risen. A rather different impression is given when comparing their contribution to the wealth of the country. While all sectors showed some growth in value of production between 1980 and 1984, the contribution of agriculture was less than half that of mining and manufacturing. What the table does not show is that some twenty years earlier agriculture contributed about 40 per cent and manufacturing only 15 per cent of the wealth. Quite obviously the economic 'miracle' has been greatest in the manufacturing sector.

It is also interesting to see the very high proportion of people employed in the services sector, which is unusual in a developing Third World country. The low and falling level of unemployment would be the envy of many industrialised nations.

The growth of world exports

This growth in manufacturing was based on a determined effort to sell manufactured goods to other countries. At first there was a concentration on light manufactures such as textiles and footwear. This became more diversified to include first heavy industry with its production of steel, ships and cars, and later a wide range of electronics, precision instruments and consumer goods. Another major source of foreign earnings was large-scale construction work undertaken by South Korean companies in many parts of the world.

During the 1960s and 1970s South Korean exports grew at a phenomenal rate, and the country became a threatening competitor to many of the previous leading exporters of manufactured goods such as the United Kingdom, the USA and Japan. The major items of trade and the main trading partners are shown on page 128.

Giant industrial corporations

This rapid industrial development has seen the growth of a number of huge corporations such as Huyundai, Samsung and Daewoo. Samsung was only established in 1969, but after about a dozen years it had 12 000 employees in its major plant, 2300 of whom lived on-site in dormitories provided by the company. Selling under brand-names such as Philco, Longines and Sears, the

South Korea's employment structure, 1984

Sector	Numbers employed (thousands)	Percentage of total workforce	Percentage change 1983–84
Agriculture	3 909	26.1	− 9.4
Mining and manufacturing	3 492	23.3	+ 3.2
Services	7 015	46.8	+ 2.9
Unemployed	567	3.8	− 7.5
Total	14 984	100	

The contribution of different sectors of the economy to GDP, 1984

Sector	Percentage contribution	Percentage change 1980–84
Agriculture	14.7	+ 1.6
Mining	1.4	+ 4.0
Manufacturing	30.1	+ 7.0
Construction	8.5	+ 6.3
Electricity water and gas	2.2	+ 9.8
Services	43.1	+ 5.2
	100	+ 5.3

company produces millions of TV sets, video-recorders, personal computers, washing machines, microwave ovens, stereo-cassettes and so on. With a new colour TV factory in Portugal it is becoming a multinational company. Some of the many products from the huge and successful Huyundai company are suggested in the advertisement. It owns one of the largest shipyards in the world in southern South Korea. In the modern yard with its many dry docks up to forty ships can be built at the same time. It takes the yard little more than a year to finish a ship, with the employees building everything from the hull, superstructure and funnels to engines, propellers, radar sets and furnishings. Their claim is 'We are just a bit faster and a lot cheaper than the Japanese.'

A part of this economic success is due to the outside aid described earlier, but most is due to the hard work and skill of managers and employees and their determination to do well. A western journalist in a recent article wrote:

'They are very regimented and I asked my guide why he, a bright intelligent young man, literate, foreign speaking, agreed to the strict regimentation his company imposes on him. "They are very good to me here. I have a good bed in a nice dormitory. They give me socks – see these? – and underwear, and a big bonus at Christmas. They look after me well, and they do my country good. No strikes, no unions, no unhappiness. All working for the good of our country – that's right, don't you think?" That was a persistent theme, and one with which I could not, after a few days' witness at the miracle, find much fault.'

Sunday Times magazine,
17 November, 1985.

The South Korean economy, like all others, has its difficult times. The government wants to see more smaller companies, for example, with a greater emphasis on Korean research and development. Construction companies operating abroad have been less successful in the middle 1980s, and the country

remains very vulnerable to the changing prices of oil imports. But the general trend is still towards industrial growth, with the future bright.

South Korean cars at the dockside waiting to be exported. These were destined for the USA

1 List some of the products manufactured by the Samsung and Huyundai companies that are in direct competition with UK manufacturers.
2 What are some of the reasons given for the success of South Korean industry?

An advertisement for the Hyundai Corporation. 'Hyundai' is the Korean word for 'modern'

THE HYUNDAI STORY

In 1975 the first Hyundai car rolled off the assembly line.

Now, eleven years on, a new highly automated plant is on stream with a production capability of up to 450,000 units a year for export to 60 countries.

Such is the Hyundai reputation for exceptional motoring value that in Canada alone, after only 2 years the marque has already overtaken Honda, Nissan and Toyota in overall sales.

When the Hyundai Stellar was launched in the UK no less an authority than the Financial Times called it 'The bargain of the half century.'

And the new Pony 5-door is giving the competition a run for considerably less money. Small wonder Hyundai is known as the rising car company.

But its star is also in the ascendency in almost every industry you care to name.

In the USA Hyundai stands for everything from micro electronics to modern furniture.

Offshore, Hyundai oil rigs pump the rich resources of the North Sea, the Persian Gulf and the Gulf of Mexico.

Since 1974 Hyundai yards have built over 200 ships including cargo and container ships, tankers and super tankers for countries from Sweden to Brazil.

The Hyundai badge is on buses, trucks, materials handling equipment, even lifeboats.

Hyundai railways and bridges are establishing new communication links as far afield as Africa and the Far East.

Hyundai housing, hospitals and schools are improving living standards in many third world countries.

In Korean 'Hyundai' means modern. For the world it spells prosperity.

HYUNDAI

3 Relations with other countries

International trade and business

The spectacular improvement in economic production in South Korea has depended on world trade. The country has needed to import many raw materials, and sell many manufactured products abroad. It has also been necessary to import vast amounts of oil to provide the energy for economic activity. The economic miracle would not have been possible without this global trade. The main countries and products involved in this trade with South Korea are shown in the diagrams. The country is also deeply involved in the world of international finance. Huge sums of money have been borrowed from overseas banks and countries during the past few decades, and like other debtor countries South Korea has suffered from big increases in interest rates. Unlike most, though, South Korea is not considered a worrying financial risk by the banks and governments that lent the money. There is a general confidence that because of the economic success, South Korea will be able to repay the debts and the interest without too many problems.

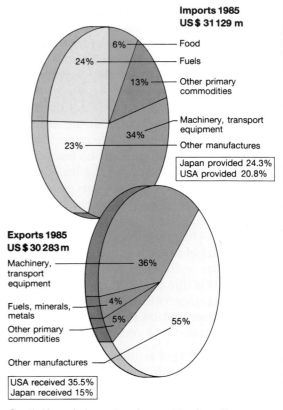

Imports 1985
US $ 31 129 m

- 6% — Food
- Fuels
- 13% — Other primary commodities
- Machinery, transport equipment
- 34% — Other manufactures
- 24%
- 23%

Japan provided 24.3%
USA provided 20.8%

Exports 1985
US $ 30 283 m

- Machinery, transport equipment — 36%
- Fuels, minerals, metals — 4%
- Other primary commodities — 5%
- Other manufactures — 55%

USA received 35.5%
Japan received 15%

South Korea's import and export trade pattern

Links with Japan

There has been contact between Korea and Japan for centuries, and in the past the tribes that eventually formed Japan absorbed much from the Korean peoples through their craftsmen and scholars. Relations changed dramatically in 1910, when industrialised Japan annexed Korea. Thousands of Koreans were sent to Japan to work in menial jobs and were compelled to take Japanese names, but were not allowed to become Japanese citizens. Korea effectively became a Japanese colony.

Since the defeat of Japan in the Second World War and the gaining of independence in 1945 relations have been strained by Japan's enormous economic influence and South Korea's position as a sort of trading or economic colony. In the mid-1980s, for example, South Korean exports to Japan were only a little over half the value of imports from that country. It was not until some twenty years after diplomatic, trade and business links were established between the two that their leaders visited each other's country, in spite of

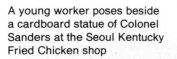
A young worker poses beside a cardboard statue of Colonel Sanders at the Seoul Kentucky Fried Chicken shop

also being close neighbours. South Koreans want the Japanese to officially apologise for their behaviour during the colonial period and the Second World War, while many Japanese feel that this has been effectively done and resent these continued accusations. The leaders of both countries want to see better relations, but many groups, including the North Koreans, are less keen.

Relations between North and South Korea

Although some meetings of officials have taken place, there have been no meetings between the heads of North and South Korea since the division of the country into two parts over forty years ago. President Kim of North Korea has vowed to reunify the peninsula in his lifetime, and South Korea lives in perpetual fear of an invasion from the north or other forms of infiltration and disruption within its borders by the communist North Koreans or their supporters.

The frontier and demilitarised zone between North and South Korea is only about forty miles and less than an hour's drive from the large capital city of Seoul. All along the roads to the northern frontier there are anti-tank barriers and military defences. There are 45 000 American troops in the country, with spy planes constantly monitoring the heavily mined and defended frontier. This is one aspect of general American support for the anti-communist regime in South Korea, and American influence is widespread. The North Koreans see this presence of such a large American force as a major cause of tension between the two countries. Another is the continued division of families. It is estimated that about 10 million people have relatives living in either North or South that they are unable to see or rejoin because of the political mistrust and hatred between the two countries. Many attempts to arrange reunions of families have failed, and only a minute number of families have been reunited in these past forty years. Many divided families do not even know if their relatives are alive or dead.

The fear of invasion, not without reason since it has been threatened often

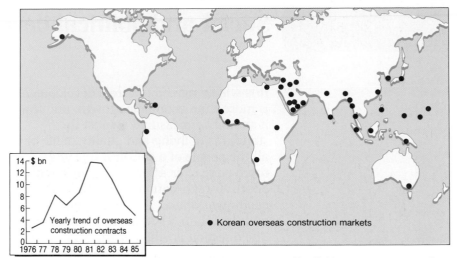

Yearly trend of overseas construction contracts

● Korean overseas construction markets

South Korea overseas: major construction operations. These have been a major source of foreign earnings, although their value has fallen during the 1980s (see graph, inset)

enough, and the assassinations of cabinet leaders while abroad and terrorism within the country, lead to vast amounts of money being spent on the army and police and all the equipment and constructions thought to be needed for defence. It also leads to strict control over any forms of protest about the actions of the government, army and police.

1 What are some of the ways in which America and Japan can be said to be nowadays 'colonising' South Korea? What name is sometimes given to these newer forms of control of one country by another or others?
2 Why is America willing to give so much military aid to South Korea, and why is North Korea anxious about recent friendly relations between South Korea and the USSR?

American troops in South Korea near the Demilitarised Zone

4 Social and political development

It is widely agreed that real development is more than economic growth, and that it should be measured against the standards of living and quality of life of all the people of a country. In 1984 there were over 40 million people with South Korean nationality. The GNP per capita was almost $2000 compared with less than $100 in the early 1960s. The question is whether this increased wealth has been widely distributed amongst the population.

In terms of health it looks as though everyone has benefited, with the life expectancy reaching 65 years. Between 1960 and 1984 the death-rate dropped from 13 per thousand to 7 per thousand inhabitants. Because of a persuasive family planning programme, a trend towards later marriage and a generally higher standard of living, the birth-rate also fell during the same period from 43 per thousand to 23 per thousand. For

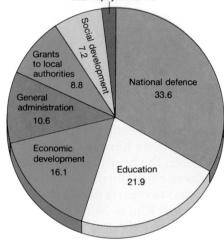

Percentage use of GDP by the government (1985)

Loan repayments 1.8
Social development 7.2
Grants to local authorities 8.8
General administration 10.6
Economic development 16.1
National defence 33.6
Education 21.9

How the Korean government spends its money

these combined reasons the total population growth dropped from about 3 per cent to 1.5 per cent per annum, and is continuing to fall. As with China, the aim is to reduce it to 1 per cent growth. The population of South Korea is very young, with over 55 per cent being under 25 years of age. Literacy has also improved, and in 1984 stood at about 93 per cent. All these signs suggest that standards of living have improved for everyone, but official statistics show that during the 1970s the share of national income received by the wealthiest 20 per cent rose from 41 to 45 per cent, that of the poorest fell from 19.6 to 16.1 per cent. So while all may be better off, the share of wealth is more uneven than it was.

Another feature of recent population change has been urbanisation, with a higher proportion of the total population living in towns and cities, and some of these growing to great sizes. Seoul is rapidly approaching a population of 10 million, putting it in the group of ten largest cities in the world. The population growth there, made up of natural increase and migration into the city, is well over 3 per cent per annum. Pusan, the second city and large port on the opposite side of the country, has a population of over 3 million.

Seoul is one of the largest cities in the world

Contrasts between rural and urban living standards

By the late 1960s an alarming gap was beginning to appear between rural and urban incomes, contributing to the migration from countryside to town. This not only puts yet more pressure on housing and other facilities in the cities, but also leads to the loss of the more energetic and educated young people from the countryside. To try and counteract this the government sponsored the 'New Village Movement', a self-help scheme which encouraged rural communities to undertake improvements to local facilities such as road building, electrification and water supply, and to develop local industries such as processing farm produce. The government provided advice, training and raw materials, and the local people the labour and leadership. The programme was successful enough to be tried in some urban areas, and it is now called the 'New Community Movement'. These programmes are an attempt to spread the benefits of growth more widely.

Personal freedoms and human rights

Partly because of recent history, and partly because of the continuing threat from other countries, the army has a very big influence in South Korean affairs. It effectively rules the country through the Democratic Justice Party and the politicians it supports. The press and broadcasting are rigidly controlled, and are used to support the government and President Chung. Student opposition to the government is put down with harsh force. Education is prized in the country and there are large universities, and these are often the centres of opposition. An article in *The Sunday Times* commented:

> 'The ruling party insists that students are the agents of Kim Il Sung (leader of North Korea), and will bring chaos and communism to the pleasant cities of the South … almost every day militants, radicals and troublemakers are hauled in for questioning, or worse.'

There is a long tradition of the Confucian religion among the Koreans, and it is sometimes claimed that this makes them willing to accept authority, and to do *en masse* what they are told is right for the nation. In recent times that authority has rested mainly with the army, the Americans and their Korean supporters. It is hard for people other than the South Koreans themselves to say whether these strict controls and limits to personal freedom are a reasonable price to pay for increasing economic wealth.

1 Draw graphs to show the population data given in this section. How do they suggest South Korea is leaving the group of Third World countries?
2 Lee Yul-Sok, a scholar-poet, wrote 'Whether a nation allows freedom of expression determines whether it founders or prospers'. Explain what you think is right and what is wrong about this view.

Every afternoon, at five o'clock, everything and everyone comes to a standstill in South Korea as the national anthem is played

Riot police stand by while South Korean students demonstrate

INDONESIA
1 Environment and history

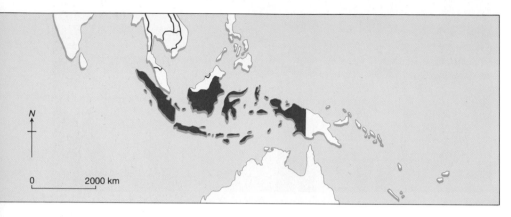

For large-scale map, see page 117

A market scene, Prapat, Sumatra

The Republic of Indonesia has a remarkable shape. It is made up of the three large islands of Sumatra, Java and Sulawesi, dozens of medium-sized ones such as Bali, Flores and Timor, and some 13 000 smaller ones. It also includes the southern part of Borneo, known as Kalimantan. The western half of New Guinea was forcibly claimed by Indonesia in 1969, and is now known as Irian Jaya. The eastern half of Timor was occupied in 1975 and made a province in 1976. These islands and their surrounding seas occupy a huge area almost as large as Europe, straddling the equator in the south-west Pacific. To the north are the neighbouring countries of Malaysia and the Philippines, to the east is Papua New Guinea, and a short distance to the south across the Timor Sea is the huge and fairly empty country of Australia.

Physical environment

Much of Indonesia consists of high mountains, mostly cloaked in dense tropical rainforest. In Kalimantan these mountain ranges are in the central and north-western parts, where they extend into Malaysia and Brunei. Java, Sumatra and most of the islands have long spines of steep-sided mountain ranges, forested and cut into by swiftly-flowing torrents. Some of these peaks reach to well over 3500 metres (11 500 feet), approaching the height of many Alpine peaks. The whole region is a zone of frequent earthquakes, and active volcanoes are found on many of the islands.

Below the rugged mountains lie more gentle foothills and fringing coastal plains. Most of the lower land was forested and swampy, but in many places the rainforest has been cleared and the land drained to make farmland. Where there has been a lot of volcanic activity the soils are generally fertile, but elsewhere the clearance of forest reveals soils that are thin, acidic and very easily exhausted and eroded when cultivated. Indonesia has an equatorial climate, and is hot and humid all the year round. Over most of the islands rainfall is plentiful throughout the year, although the monsoon rains between about October and March are much heavier than during other months. The islands extend between latitudes 6° North and 11° South, and this means that the islands in the south-east, nearest Australia, do get a drier season. It is this equatorial climate that has produced the widespread rainforests in the interior and the mangrove forests and swamps along the coasts.

The people and their history

Over 164 million people lived in this vast sprawl of islands in the mid-1980s, making it the fifth most heavily populated country in the world. If the population continues to increase at the rate predicted the total will reach 330 million by the year 2050, making Indonesia the third most populous country.

Traces of the first inhabitants go back more than 30 000 years, and there are reliable records of events over the past several thousand years. From the seventh century onwards a Buddhist empire flourished in Sumatra, and later a Hindu-based empire existed in Java for over three centuries. The influence of Islam began to be felt due to the activities of traders almost a thousand years ago, and eventually Islam became the dominant religion. Indonesia is the largest Islamic state in the world, with a large minority of Christians and smaller proportions of Hindus and Buddhists. The people of Indonesia are a great mixture of different ethnic, cultural and religious groups.

In 1511 the Portuguese took possession of Malacca in Sumatra, and tried to establish an influential spice trade over the whole area. They failed, but the Dutch who succeeded them were more successful. The Dutch East India Company was founded in 1602, and from then on for almost 350 years the islands were in effect an economic and political colony of the Dutch. Plantations were established and much of the produce was shipped to the Netherlands and elsewhere in Europe, as were various raw materials. In return goods manufactured in the Netherlands were shipped out to Indonesia. The merchants and people of the Netherlands profited greatly from this colonial trade, as did some Indonesian merchants, farmers and rulers, but on the whole the bulk of the people gained very little of the wealth created.

As in most colonies, the twentieth century saw the emergence of nationalist movements in Indonesia. The Indonesian people wanted to run their own affairs and receive what they felt to be their fair share of wealth produced in the country. The Japanese took control of the islands during the Second World

War, but the Dutch returned after the defeat of the Japanese in 1945. They found great difficulty in retaining control by force, and finally Indonesia gained its independence from the Dutch in 1949.

1   What are some of the advantages and some of the disadvantages of being a colony of a European power over several centuries?
2   What are some of the problems facing anyone wanting to create a nation-state of Indonesia?

These beautiful and ancient rice terraces on the island of Bali are a fine example of integration between people and their environment

Jakarta, the capital of Indonesia, is one of the largest and most densely populated cities in the world

INDONESIA

2 Economic development

With its high temperatures and reliable rainfall all year round lowland Indonesia is well suited to agriculture. Over half the working population is engaged in farming, a far bigger proportion than any other type of occupation. Farming provides only about a quarter of the national income, however. A great deal of farm production is from small peasant farms and smallholdings. Rice is the most important food crop, and this is mainly grown on the lowland paddy fields. Other food crops include maize, cassava, sugar-cane, sweet potatoes and a wide variety of fruits and vegetables. Local fish production is another important source of food. Farming has become more efficient and productive, but Indonesia cannot grow enough rice for its large and growing population and some must be imported.

Very different from these small farms are the much larger plantations that grow crops of tea, coffee, coconuts, oil-palm, nuts and rubber. Many of these estates were created by the Dutch, and from the beginning were meant to provide cash crops for export. Other large tracts of the rainforest have been cut for timber. Partly due to a failure to replant after cutting, and partly due to growing world concern about the destruction of the rainforests, there has been a decline in the rate of felling and exporting of timber.

Minerals provide an important part of Indonesia's wealth. There is plenty of oil and natural gas. Crude and refined oil and liquid natural gas provide almost three-quarters of export earnings. Coal, tin and nickel ores are mined, while there are large and untapped reserves of other minerals. Manufacturing is not a very important part of the economy, and manufactured products make up only a small proportion of exports. There are factories making products such as textiles, fertilizers, cement, paper and household goods mainly for the home market. Many manufactured goods have to be imported, which is quite usual in Third World countries.

Recent development

Indonesia has been an independent country for almost forty years. It has had to cope with many difficulties in trying to develop its economy. There are good air and sea links between the islands, but the wide scatter of separate parts that make up the country is a disadvantage. Many people are educated and skilled, but large numbers are poor, unskilled in modern methods of work and illiterate. Indonesia has achieved a steady economic growth since independence, but as will be seen, this has not been matched by an even distribution of the wealth created.

The pattern of South-East Asian trade

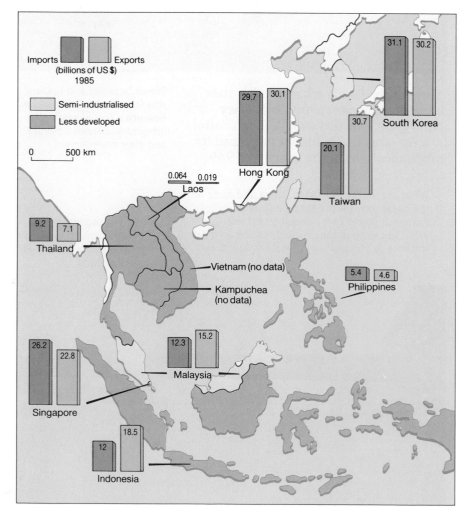

The dual economy

Indonesia shows a pattern of growth common to most Third World countries. On the one hand there is a foreign-owned export sector involving the plantations, mines, oilfields and larger factories. These produce commodities for the factories and processing plants of the First World. They are capital-intensive, needing lots of money to get started. The country gets some wealth from these activities, as do the foreign investors. A relatively few Indonesian businessmen, professionals and skilled workers also earn enough to live in comparative wealth.

On the other hand the majority of Indonesians are hardly touched by this side of the economy. As farmers, or as workers in the manufacturing, mining, construction, transport or service industries, they either get very modest wages or survive by producing their own basic foods and other needs. As in most Third World countries there is a great deal of underemployment and unemployment, with the usual evidence of poverty.

The basic aim of the government has been to use the massive foreign aid and the foreign commercial investment to develop the natural resources of the country. Money earned by the sale of these products abroad would pay for imports and further development. Unfortunately this economic policy has had to face a number of problems. The price of oil, so vital to Indonesia, dropped dramatically during the early 1980s, and experts have predicted that reserves might run out by the end of the decade. There is little chance of a manufacturing export boom to take its place, and the income from the exports of timber and minerals has declined. Indonesia still relies heavily on foreign aid, and there is no guarantee that this will continue forever. The third Five-Year Economic Plan (1979–1984) was designed to encourage 'the equitable distribution of the fruits of development' at the very time when the means of doing so were getting much more difficult.

This poster depicts social and economic progress in Indonesia, and urges the Indonesian people to support the government's latest development plan

Jakarta sets up team to spend aid faster

Indonesia has set up a special ministerial team to solve a pressing problem – how to spend aid money more quickly.

This comes after complaints from aid donors and the World Bank about slow implementation of development projects in 1984–85.

Jakarta has blamed the delays on land acquisition problems, budgeting and finance procedures, finding suitable subcontractors, poor management and red tape.

Aid projects include new rubber plantations, dams, roads, housing, resettlement and electrification schemes.

1 In what way is the economic structure of Indonesia typical of a Third World country?
2 Why was foreign aid being used so slowly in 1984–85? What was aid money being used for during that time?

Indonesia's import and export trade pattern

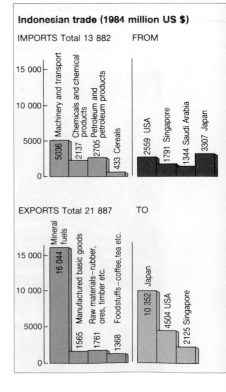

135

3 Sharing in development

In 1982 the World Bank announced that Indonesia had become a middle-income developing country, with its steady rise in GNP per capita. At the same time, however, several changes were suggesting a bleaker future. Oil exports were getting lower and world prices falling. Indonesia is a leading producer of rubber, tin, palm-oil and tropical hardwoods, but these commodities are subject to rapid changes in demand and price. In fact, commodity export earnings fell by about 40 per cent in the year before the World Bank announcement. It is easy to see why the government is keen to move away from dependence on the export of oil, gas and basic commodities. In spite of these difficulties, though, Indonesians should have been benefiting from the increasing wealth of the country.

Deciding how best to use foreign earnings and foreign aid is not easy. There is a big need for many new jobs, but much of the earned income, as well as foreign investment, has been put into capital-intensive rather than labour-intensive industries. New laws have been introduced to encourage local Indonesian businesses to get started, but some claim the main reason for these is to curb the influence of Chinese In-

donesians. They represent only 2 per cent of the population, but control over 50 per cent of private industry.

Poverty and the 'lower economy'

The people who will suffer most if the problems of economic development are not solved are those who are already the poorest.

Several factors make the improvement of living standards for the poorer members of society more difficult. There is a very uneven distribution of land ownership, for example. Of the 120 million or so who live in the countryside about 40 million are without any land, while another 46 million own less than a quarter of a hectare. Moderate land reforms were suggested in 1960, when it was proposed that land would be made

The 'lower economy' in urban and rural areas: inside a Manila cigarette factory (below), and winnowing rice in the traditional manner (right)

Development expenditure (billion rupiahs), 1985–86	
Agriculture/Irrigation	1 430
Industry	655
Mining/Energy	1 301
Tourism/Communication	1 425
Manpower/Transmigration	676
Regional, rural and urban development	868
Education	1 510
Health, social welfare, womens affairs, family planning	413
Housing, Settlements	437
Defence and Security	714

Total expenditure		Total income	
Development	10 647	Earned domestic income	18 677
Other	12 400	Foreign aid	4 368

Refugees from Irian Jaya at a camp in Papua New Guinea

available to those prepared to cultivate it. The reforms were never carried through, and most proposals for land reform are opposed because they are seen as being a way of introducing communist ideas and practices into the country.

Government credit is available for some types of development but cannot be claimed by many landless peasants and small landowners. The poor have to rely on private money-lenders who are able to charge interest at up to ten times the government rate. Middlemen are also able to control trade in rural areas, and pay low prices for farm produce but charge high prices for city-produced products. Such middlemen are often in collaboration with local politicians and officials. In Indonesia, as in most parts of the world, real development and the hoped-for distribution of wealth is hindered by the greed and corruption of some businessmen, officials and politicians who have power and influence.

East Timor and West Irian

East Timor was for centuries a Portuguese colony, but in 1974 the newly-elected government of Portugal withdrew from the former colonies. Following armed struggle between several Timorese political parties, the strongest emerged and declared East Timor an independent nation in 1975. Indonesia was unwilling to accept the situation and invaded the eastern part of the island. The United Nations recognises the right of East Timorese to 'self-determination, freedom and independence', but this is ignored by the Indonesians. The resistance movement is now confined to the remote mountain areas, but the conflict continues to make economic and social development more difficult.

The western half of the island of New Guinea was the only area not handed over to Indonesia by the Dutch when their colonies gained independence in 1949. It was agreed that the peoples of western New Guinea would determine their own future. When it seemed this would not happen, Irianese leaders took up arms, and in 1966 began fighting for a Free West Papua. Indonesia arranged consultations with some limited groups, claimed their approval, and annexed West Irian in 1969. Armed resistance continues in Irian Jaya, adding to the problems of development and complicating relationships with neighbouring Papua New Guinea.

This child is one of the lucky few to have access to a public health centre

1 Look at the photograph of the cigarette factory. What does it show about manufacturing methods in the Third World?
2 Apart from East Timor and Irian Jaya, describe some of the reasons why the Indonesians find it hard to improve the living standards of all the people.

4 Migration within Indonesia

The complicated shape of Indonesia, with its scatter of many islands, has already been stressed. What is not quite so obvious is the remarkable variation in the density of population from one part to another. Java and the neighbouring islands of Bali, Madura and Lombok are seriously overcrowded and all suitable land is already cultivated. By comparison Kalimantan and most of the outer islands have a very low density of population. Kalimantan is almost four times the size of Java, but it has only one-fifteenth of the population. These imbalances between land, resources and people result in over-population and under-population in different parts of the country.

People have always migrated within Indonesia, but in the late 1970s the government announced a plan to transfer masses of people from the overcrowded parts to the sparsely populated and undeveloped outer islands. During the first five years of the programme, between 1979 and 1984, the aim was to move 500 000 house-holds, or about 2.5 million men, women and children, into previously unsettled areas. Such a massive migration placed enormous demands on the government and thousands of officials.

There were other goals as well as releasing population pressure on the central islands. The intention was to create new farmland and it was hoped that the migrants would produce badly-needed food and additional export crops such as rubber and oil-palm. There had also been some earlier development of mining and manufacturing in a few new centres scattered around the islands. Natural gas is liquefied in Kalimantan and aluminium manufactured in north-ern Sumatra, for example. Development of the outer islands included providing more facilities for the peoples of these new industrial centres. Transmigration was seen as just one of the activities to develop the outer islands.

The Indonesian population distribution

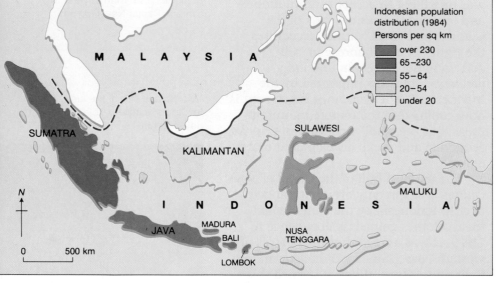

Indonesian population distribution (1984)

Persons per sq km

- over 230
- 65–230
- 55–64
- 20–54
- under 20

This overcrowded, insanitary slum in Jakarta is in contrast to the new settlements to which the people (facing page) have migrated

Problems of the transmigration programme

One of the striking facts about the programme is that even if the plan to move 2.5 million people in five years had worked, the population of Java alone would have grown by about 9 million! In spite of considerable success in family planning programmes, the population is still increasing very rapidly. During the five years about 250 new settlements were established, including schools, community centres, co-operative services, health centres and mosques or temples. But the massive migration ran into a number of difficulties.

Not everything has gone smoothly in the new settlements. The usual pattern has been for the government to clear a patch of forest and provide shelter, basic health care and schooling. The new families move in and are given help to see them through the first year or so until the crops they plant are harvested.

Settlers often have to cope with local soil and weather conditions which are very unfamiliar. Yields of crops are often low while the farmers get used to their new environment. Because income is low it is hard for newly-settled farmers to buy sufficient seeds or fertilizers and to develop much beyond subsistence level. Even so the average family income is higher than previously, which is not surprising since the majority of settlers are from the poorest rural areas or are the urban homeless.

Lack of local knowledge and necessary farming skills is only one handicap. Farms are frequently too large to cultivate without draught animals, and there are not enough of them for everyone. The government has experimented with providing small tractors to co-operatives for renting out to settlers. Farmland made out of cleared forest is vulnerable to attack from the wild animals of the surrounding forests, and it is thought that as much as a third of the produce can be lost in this way. The new villages are also often short of craftsmen and other skilled workers.

Some villages have experienced problems because of the great variation of religion, language and culture amongst the newcomers. One answer

has been to try and settle 25–30 households from the same area in the new blocks of land. Another difficulty has been the resentment shown by some local people towards the newcomers, not only because of added pressure on the land but also because they usually lack even the basic facilities the new settlers get as part of their aid package on moving. As a result a percentage of newly-developed land is reserved for local people. The main hope is that after the settlements have become fully established the transmigration programme can end and be replaced by a big increase in voluntary migration. In the meantime the huge resettlement scheme continues.

1 What would seem to be the likely advantages and disadvantages of resettlement from, say, Jakarta or a densely settled farming area for a fairly poor household?
2 What are the costs and benefits of the transmigration programme to the country of Indonesia as a whole?

This resettled family has been provided with a house and the equipment shown, but often this is not enough. Many find making a living a permanent struggle

A resettled farmer poses in a newly-planted rice field in recently-cleared forest. In the background is the new village. But clearance of the forest often leads to soil erosion, and many families are living on sites that are incapable of sustaining them

139

SRI LANKA
1 Environment and history

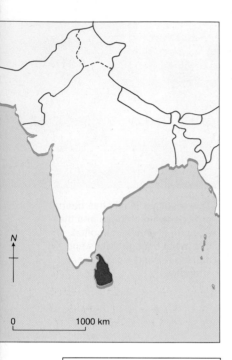

For large-scale map, see page 118

The Republic of Sri Lanka is a pear-shaped island lying some 35 kilometres off the south-east tip of India in South Asia. Over 70 per cent of the population are Sinhalese, and most of these are Buddhists. Another large group are the Tamils who make up about 20 per cent of the total. The majority of the Tamils are Hindus. Until it gained its independence Sri Lanka was a British colony and was known as Ceylon.

Physical environment

Sri Lanka is about 440 kilometres (275 miles) long, and 220 kilometres (138 miles) wide. The map shows its size and shape compared with the British Isles. The centre is mountainous, with the highest peak over 2500 metres. This is well over 1000 metres higher than Ben Nevis, the highest mountain in Britain. There are many peaks, ridges and escarpments in the central mountainous area. Lower foothills lead from this mountain area to the surrounding coastal plain. This coastal plain is narrowest in the south-west.

Some idea of the general lowland climate can be got from the climate

This is what the tourists come for; a palm-fringed beach

graphs. Most of Sri Lanka gets enough rainfall for crops to grow throughout the year, although there is a much drier season between March and June in the north-east. The effect of this is more marked because much of the island here is underlain with limestone rocks through which the water easily drains. Temperatures are fairly uniform and high all the year round, and so there is no cold season to adversely affect farming. Because the heavy monsoon rains of May to October come from the south-west, it is that part of the island and the mountains that receive the heaviest falls. Many rivers fed by this rainfall drain out of the central mountain areas. They are being used increasingly to provide irrigation water and hydro-electric power. Much of the original lowland jungle has been cleared, but areas of tropical rainforest remain in the wetter areas. Much of the cleared land is used for growing crops on farms and large plantations, and it is these that have formed the basis of the country's wealth up till now.

The physical environment of Sri Lanka

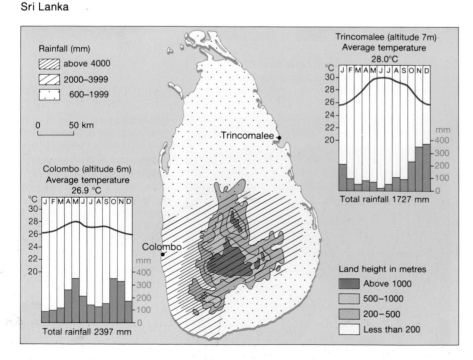

Rainfall (mm)
- above 4000
- 2000–3999
- 600–1999

0   50 km

Colombo (altitude 6m)
Average temperature
26.9 °C
Total rainfall 2397 mm

Trincomalee (altitude 7m)
Average temperature
28.0°C
Total rainfall 1727 mm

Land height in metres
- Above 1000
- 500–1000
- 200–500
- Less than 200

Tourism is an important industry in Sri Lanka. This is typical of one of the more expensive hotels

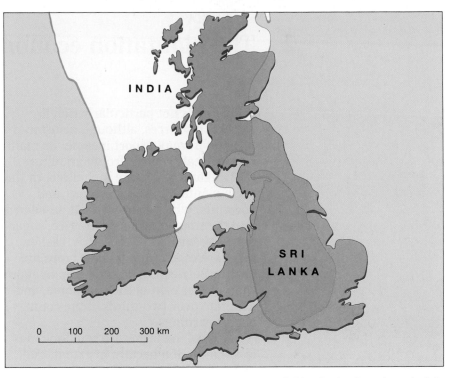

Sri Lanka (and southern India) compared with the British Isles

Historical background

There is evidence of human settlement on the island going back for thousands of years, but the place of origin of these first people is unclear. About 2500 years ago, during the fifth century BC, Indians from the mainland are known to have settled along the west coast. Over the centuries they moved inland, and established their territory in the north-west part of the uplands with a capital at Anuradhapura. As early as the fourth century BC they had adopted the Buddhist religion, and their culture lasted for over a thousand years. During this time they developed very efficient methods of farming and many irrigation schemes.

Towards the end of this period many Indian Tamils settled in the north of the island and built their capital at Jaffna. And so the present-day pattern of settlement began to develop, with Sinhalese in the centre, west and south of the island and Tamils in the north and east. Between the sixteenth and eighteenth centuries first Portuguese, then Dutch and finally British traders took possession of the island, much as they took many other parts of the world. Ceylon, as it was then called, became a British colony in 1802. As was often the case the Europeans established large plantations – in this case of coffee, tea, rubber and coconuts – to provide food and raw materials for the homeland. To work the plantations more Tamil labourers were brought in from India

during the nineteenth century, and they now form a second Tamil minority in the central highlands.

Ceylon was given partial independence in 1947, and became a fully independent nation with a new name of Sri Lanka in 1972. It became a member of the Commonwealth, just one of many reminders of its previous control by Britain. The present pattern of population, types of economic activity, social and cultural features, and the violent conflict between Sinhalese and Tamils, reflect both the physical environment and the past history of the island.

1 Describe some of the environmental and cultural contrasts within Sri Lanka.
2 What do you think is likely to a) favour and b) hold back economic and social development in Sri Lanka?

A demonstration by Buddhist monks in favour of peasants' rights

2 The plantation economy

Sri Lanka is not particularly rich in mineral resources, although gemstones bring in a small export income. Its main export wealth comes from agriculture. Sri Lankan agriculture includes both the huge plantations of tea, rubber and coconuts, which are largely state-owned, and the smaller farms where rice, sugar-cane, cassava, sweet potatoes, cashew nuts, spices and many other crops are grown. Between them they employ more than 40 per cent of the workforce, and generate over two-thirds of the country's export earnings.

Most of the farms are smallholdings less than half a hectare in extent, and are run at little more than subsistence level. The island is not self-sufficient in food production, however, and needs to sell its plantation crops to pay for food and other imports. Sri Lanka's economy is very dependent on these plantation crops.

Copra, the dried husks of coconuts, are a major export product

The tea estates

Tea is the single most valuable export, and contributed about a third of the export earnings in the mid-1980s. In fact this was a considerable decline from twenty years earlier when the share was over half. During that period the total production of tea also fell due to a number of problems facing the industry.

The first tea estates were established in the middle of the last century. The first commercial tea crop in what was then Ceylon was planted by the Englishman James Taylor in 1866. The climate in the central highlands area allows the crop to grow throughout the year, and so encourages commercial production. Large areas of forest were cleared as the tea companies established their estates. Similar developments occurred in other parts of South-East Asia and later in East Africa.

When it became clear that Ceylon would gain its independence from colonial rule the tea companies realised that the tea estates would almost certainly be nationalised. This threat discouraged European owners from investing in their estates, and older tea bushes were not replaced to guarantee a sound future output. Expensive fertilizers were not added to the soil, and the estates were generally run-down by the time they were nationalised by the new Sri Lankan government. The new managers were also still gaining in experience and learning to run the estates efficiently. The total effect was a decline in the output and quality of tea. Similar problems were experienced in the coconut and rubber plantations. Soon after nationalisation tea bushes began to be replanted, both on the large plantations and on the thousands of small plots created by land reform and the division of the larger estates in 1972.

The tea industry continues to face a number of serious economic and physical problems. The price of tea is fixed by the tea-selling companies according to world demand and financial conditions. Prices can change quickly, but the growers claim that the prices they are offered are always low so that the selling companies can maximise profit. It is hard for the growers to object, since the selling companies would take their business elsewhere. The only solution is for the growers to increase their efficiency.

The distribution of the world's tea-growing areas, and the major tea importers

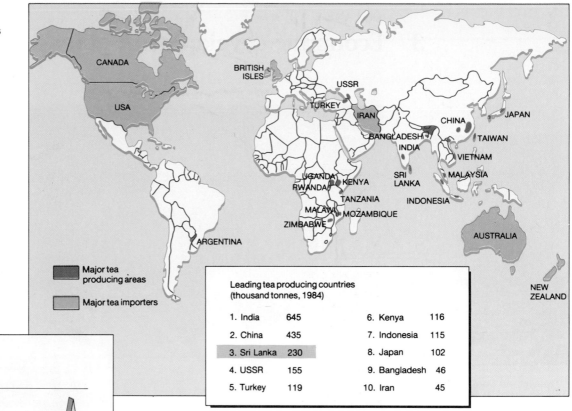

Major tea producing areas

Major tea importers

Leading tea producing countries (thousand tonnes, 1984)

1.	India	645	6.	Kenya	116
2.	China	435	7.	Indonesia	115
3.	Sri Lanka	230	8.	Japan	102
4.	USSR	155	9.	Bangladesh	46
5.	Turkey	119	10.	Iran	45

The table (inset) shows the leading tea producing countries. The graph shows the fluctuating price of tea

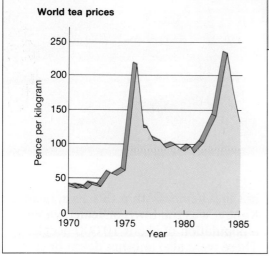

World tea prices

Pence per kilogram

Year

putting these ideas into practice would be enormous. It would be many years before any loans for improvements could be repaid, and modernisation depends heavily on the availability of aid from other countries.

1 List the physical and economic features that **a)** favour and **b)** handicap the tea producing industry of Sri Lanka.
2 Compare the data on world tea production provided by the map and the statistics. Comment on the information provided, particularly from the point of view of Sri Lanka.

Plantation managers are attempting to become more efficient by increasing soil fertility, developing different varieties of tea bush and using mechanical cultivation techniques. Contour ploughing, terracing, constructing drainage channels, using herbicides rather than manual weeding and planting grass cover during replanting are all methods used to improve and protect the plantation environment. Higher yields are obtained by using plants grown from cuttings rather than seeds, although the quality of the tea is slightly lower and such plants are less resistant to drought should the monsoon rains fail. The Sri Lankan Tea Research Institute is able to make many suggestions for the improvement of the tea estates, but the cost of

A tea plantation in the cool central highlands

3 Economic development: gains and losses

The Victoria Dam on the Mahaweli River, under construction (main photo) and after completion (inset). Oil is Sri Lanka's most expensive import, and the Mahaweli scheme will save costs by providing cheaper hydro-electric power

Sri Lanka's major irrigation areas, including the proposed Mahaweli River Project

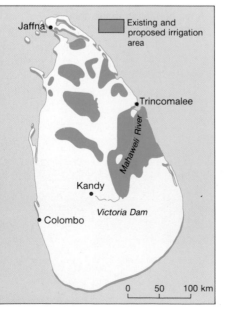

The Mahaweli Development Scheme

One of the many rivers flowing from the rainy central highlands eastwards across the drier lowlands to the Indian Ocean is the Mahaweli. The Development Scheme, started in 1977, aims to triple the electricity output of this the largest of Sri Lanka's rivers. In addition irrigation water will be provided to bring more land under cultivation in the drier north-east part of the country. The scheme is also concerned with the resettlement of 1.5 million people, almost a tenth of the country's population. It would seem to offer the opportunity of creating more energy, growing more food and providing many Sri Lankans with a higher standard of living. But the development has aroused a great deal of criticism, and a number of unfortunate problems have arisen.

One major cause of bitter complaint is the drowning of a large, long-settled and fairly prosperous area under waters created by the Victoria Dam, the largest in the scheme. Within this area, near to Kandy, was the town of Teldeniya with a population of about 10 000 people. There were also another dozen or so smaller towns and over 120 villages with schools and Buddhist temples. All these people were resettled in a district only about 70 kilometres away to the east, but in a climatic zone that was much hotter, but drier and less humid.

In a largely undeveloped area, with few roads or other facilities, the settlers were given an acre of land for cultivation, which was more than most had previously, and another small plot for a house. Unfortunately the rainforest and scrub had to be cleared and the houses constructed by the settlers with a minimum of equipment and help. Villages and facilities were to be provided by the government, but at a later stage. The benefits were felt to be small, and there was resentment that there had been no consultation or chance to object to this massive upheaval.

Apart from the 50 000 or so settlers from the lands flooded by the reservoirs two other groups were moved to the newly-developed areas. About 100 000 people were selected from towns and communities throughout Sri Lanka for resettlement. These tended to be the unemployed and poorly housed. The third group consisted of those already living in scattered villages in the area. Resettlement began in 1981.

Another major problem has been to do with finance. The scheme was originally costed at £700 million, an enormous sum for a relatively poor country. Britain contributed about £100 million, largely to British firms involved in the project such as Sir Alexander Gibbs & Partners, the dam contractors, the Balfour Beatty Nuttall consortium, and Costain International. Other help was given by Sweden, Canada, West Germany, Kuwait and the EEC. In the original plans Sri Lanka was left to find about £300 million, but between 1977 and 1983 the cost of the project trebled to £2000 million, and the Sri Lankans had to find more than five times their original contribution. The country has had to do this by raising taxes, cutting subsidies and services to people, and reducing support for the resettlement areas. Many people have in effect been penalised rather than helped by the ambitious scheme.

Developing industry

In an attempt to avoid dependency on the export of plantation crops, the Sri Lankan government has tried to encourage industrial growth. They have done this by providing financial help to local firms, but one of the more determined approaches has been by persuading foreign companies to invest in the country. An Investment Promotion Zone (IPZ) was created on a 250 hectare site adjacent to the international airport about 30 kilometres north of Colombo in 1978. It offered generous incentives such as tax exemption of up to 100 per cent for as much as 10 years, duty-free import of factory equipment and raw materials, and the availability of a cheap but well-educated workforce. During the first four years over 80 factories were

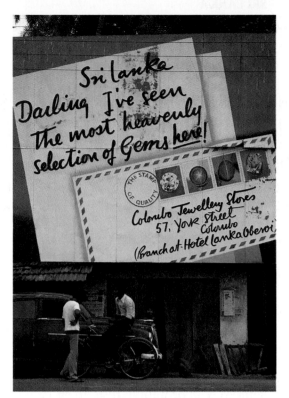

This clever advertisement draws attention to the fact that precious gems are one of Sri Lanka's major exports

established, many involved with textile manufacturing. Over 30 000 people were employed, mainly young women and girls. Quite a few foreign companies using the zone were from Hong Kong and other South-East Asian countries.

Critics argue that this has hindered Sri Lankan firms from competing in the home and overseas markets, and that the government has merely replaced an economy at the mercy of world markets with one at the mercy of foreign investors and the large multinational corporations who gain most from the scheme. There are now many consumer goods widely advertised in Sri Lanka, and life for the wealthy is relatively good. But the poorer sections of the community seem to be benefiting less from recent developments.

1 List the gains and losses caused by the Mahaweli Development Scheme.
2 What are the advantages and disadvantages of involving the people who will be affected by major economic projects in the decision-making?

4 A divided nation

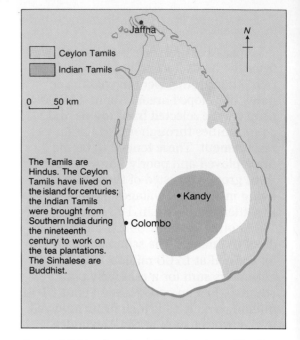

The Tamils are Hindus. The Ceylon Tamils have lived on the island for centuries; the Indian Tamils were brought from Southern India during the nineteenth century to work on the tea plantations. The Sinhalese are Buddhist.

Areas of Sri Lanka where there is a large Tamil population

The tourist industry

One of the most valuable resources of a country is a beautiful or spectacular environment. Not only can this give great satisfaction to the people of the country itself, but can be the basis of a valuable tourist industry. Tourists are also attracted by evidence of past history, and by the rich and colourful cultural traditions that may be so different from their own. When environment, past history and cultural attractions are supported by effective hotel, transport and service facilities, then the tourist industry can be very significant.

The advertisement makes a bold claim about Sri Lanka. All advertisements, probably more than other sources of information, need to be treated with some caution. Statements are made that are only a part of the truth about a place, and many less attractive features get no mention. From this advertisement Sri Lanka looks a very desirable place to visit – though other countries such as Papua New Guinea also lay claim to be 'paradise'. Many people are indeed attracted to the island, and in 1980 over 320 000 tourists visited Sri Lanka. Tourism is now the country's fourth largest foreign exchange earner. But tourism, like other development, is severely threatened by recent social tension and violence that challenge the idea that Sri Lanka is a simple paradise.

Communal tension and violence

The presence of two major groups within the island was explained earlier. The majority group are the Sinhalese. The parts of the country in which the minority Tamils live are shown on the map. In some parts the Tamils form the local majority, and in places few Sinhalese are found. The basic cause of tension between two groups is that the Sinhalese are dominant in government and public life. The Tamils feel at a disadvantage and want to establish an independent state of their own in the north and east. Quite understandably, the Sinhalese do not want the island partitioned in this way, and have resisted the proposal. Tension is made worse by the existence of a large number of Indian Tamils a short distance away across the Palk Strait who support the Sri Lankan Tamils in many ways.

The Tamils of both India and Sri Lanka reject the idea that the solution is for the Tamils to leave the island and make new homes on the mainland. This is a classic case of two groups of people within one state with different beliefs, cultural patterns and life-styles, one of

The tourist image of Sri Lanka. In the 1980s the tourist industry was badly affected by the internal conflict

Sri Lanka. Paradise.

Air Lanka. A taste of Paradise.

which feels unjustly treated and unable to improve things.

No peaceful solution has proved successful, and both sides have turned to violence. There are a number of Tamil resistance groups – or terrorist groups, according to who is describing them. There is savage disagreement amongst them about who truly represents the Tamil population, but the Tigers have emerged as the main group.

The bombings, massacres and reprisals by both Tamils and Sinhalese became worse during the early part of 1987. On 26th May government troops launched an offensive against Tamil strongholds in the Jaffna Peninsula. There was full-scale civil war until 11th June. On 30th July an Indian peace-keeping force of 3 000 troops arrived in northern Sri Lanka, and an Indo – Sri Lankan accord gave the Tamils limited self-government. But the Sinhalese then protested against this. In October Indian forces launched an offensive against the Tamil strongholds and took control of Jaffna. But, the problems remain.

It is a tragic situation that must be overcome if there is to be any real development on the island. Sri Lanka has an education and health service envied by many Third World countries, but shares with them other problems such as an economy shaped in the colonial past and a rate of population growth that threatens to cancel out any economic progress. But the end of violent conflict between Tamils and Sinhalese is the most urgent priority, especially if the reality of life in Sri Lanka is ever going to approach the picture given in the advertisement.

I am prepared to give them some degree of federalism. But that is all. If these proposals are not acceptable, I will unleash the troops.

I am having to cut down on development projects that would have benefited both Sinhalese and Tamils alike, in order to pay for this battle. Our country is being destroyed.

If I give in to the terrorists' demands, my own Sinhalese people will rise. If I refuse to give in to their demand for an autonomous state, they plant bombs, and become major world-wide terrorists, a menace to everyone. If I go hard against them, then I am accused of genocide.

President Junius Jayawardene, May 1986

Curfews and chaos on island of fear

Sri Lanka is in total chaos. If tourists had not been advised by the authorities to leave, they would be holidaying amid three-day curfews, power cuts and blackouts, fuel shortages, and fear.

Many hospitals have to use standby generators. Few buses are running, and the railway lines have been blown up in several places. The historic town of Kandy was without water and petrol. Last week, many resort hotels were forced to close. Schools have been shut for almost a month, and government offices are closed.

The government has all but declared a state of emergency, and introduced martial law. Apart from the almost permanent curfew in some areas, police and soldiers are now empowered to shoot curfew-breakers on sight and they are allowed to take away bodies and bury them without inquest.

In the north of the island, terrorists fighting for a separate Tamil homeland and keeping 50 000 Indian troops at bay last week ambushed a bus and killed 27 people.

Observer, 20th November 1988

1 Read the quote and extract, and suggest what you feel the correct course of action by the government should be.
2 List and briefly describe similar conflicts around the world, explaining whether partition has been or is likely to be the solution.

(Below left) a young Tamil stands guard over territory that he would like to see as an independent state. (Below) in 1987 units of the Indian army were invited into Sri Lanka to help solve the conflict

SAUDI ARABIA
1   Environment and history

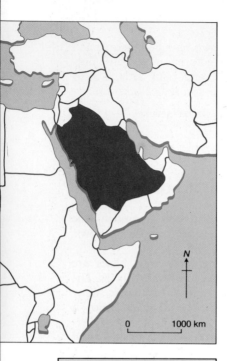

For large-scale map, see page 119

The Kingdom of Saudi Arabia consists of the central part of the Arabian peninsula between the Red Sea and the Persian Gulf. It occupies about four-fifths of the peninsula, the remainder being the states of Yemen, South Yemen, and Oman to the south, and Bahrain, Qatar and the United Arab Emirates to the east.

Physical environment

The country is very high in the west, where a steep escarpment rises from the narrow Red Sea coastal plain to over 2000 metres (6500 feet) in the north and over 3000 metres (10 000 feet) near the Yemen border in the south. From these heights the land falls in stages to sea level in the east. To the south is the great sandy desert of Rub al Khali, often known as 'The Empty Quarter' because it is so dry, harsh and barren. Summers are normally hot and dry, with winters much cooler. There is unreliable seasonal rainfall in the western mountains, where in winter there can also be falls of snow. Rivers flow from the mountains towards the Red Sea and the interior, but many cease before reaching the sea. Grasslands can be found in the mountains where there is sufficient rainfall. Most of Saudi Arabia, though, is desert, with the occasional oasis.

The emergence of the kingdom

Different tribes of Arabic people have lived in the Arabian peninsula for thousands of years. Some were cultivators inhabiting the scatter of oasis

Camels were the traditional means of transport in the desert

This barren landscape with its mixture of rocky and sandy desert is typical of Saudi Arabia

settlements, while many were nomadic camel herders who wandered the deserts. Between 1902 and 1924 a powerful tribal leader, Abdul-Aziz Ibn Saud, united the four great tribal regions of Nejd, Al-Hassa, Hejaz and Asir into a single kingdom. In 1932 he offically named this new kingdom Saudi Arabia, and that same year it became recognised internationally as an independent state.

The country is ruled by hereditary kings. Ibn Saud died in 1953, and by the mid-1980s four of his sons had succeeded to the throne and become Head of State. Many members of the royal family are ministers in the government, which is not elected as in a democratic state. The population of Saudi Arabia is over 11.2 million, and Arabic is the national language.

The influence of Islam

During the seventh century AD the Arabian peninsula became the homeland of both the Islamic religion and of Arab culture. The Islamic religion began in AD 622, the year that the prophet Muhammad and a handful of followers left the town of Mecca and travelled to the town of Medina about 200 miles to the north. He began to form his Bedouin disciples into a religious and military force. Within 30 years they had conquered and converted not only the whole Arabian peninsula but much of what is now the Middle East, the Nile Valley, and the coastlands of Egypt and Libya. Over the following centuries the Islamic faith was taken across North Africa and into Spain, into Persia, India and Indonesia, and into West Africa. Towards the end of the seventeenth century the Islamic peoples claimed power and influence over a large part of the world and superiority over the Europeans in many academic and practical skills.

In the sixteenth century Arabia had come under the rule of the Ottoman rulers based in Turkey. The Turkish Ottomans remained in control of Arabia until the end of the First World War. They had sided with Germany, and following their defeat the whole Middle East area was partitioned into British and French spheres of influence. It was

following this that the head of the family of the House of Saud took control by a mixture of conquest and diplomacy as described above.

The Koran, the holy book of Islam, provides the basis of belief and law of millions of converts, including the Saudi Arabians. It has clear and strict rules for personal and social life, and these are accepted and practiced with great thoroughness. For several decades Saudi Arabia remained a country with an economy based on pastoral farming, but then oil was discovered in vast quantities, and this has had a dramatic impact on the country and its people.

The holy pilgrimage city of Mecca from the air: a modern city with high-rise housing and traffic jams. In the middle of the photograph is the Great Mosque, with pilgrims surrounding the sacred Kaaba

1 What are the differences in the form of government of Saudi Arabia and the United Kingdom? What are the advantages and disadvantages of each?
2 Describe what is meant by nomadic pastoralism, and by oasis cultivation. Say whether you think they are typical of a more or a less developed economy, and why.

2 Food from the desert

(Right, top) improved irrigation techniques have meant that Saudi Arabia is now self-sufficient in many agricultural products. (Right bottom) much of the produce in this Saudi shop has been locally produced

Saudi Arabia, like most countries, tries to grow as much of the food and fodder crops it needs as is possible. Only a decade or so ago its food import bill was enormous, and at one time it was third in the world order of food importers. Even if there is plenty of money to pay for these imports, this reliance on others puts a country in a vulnerable position in times of crisis. The Saudis chose an alternative course, and put a great deal of money into improving agricultural production in their Third Development Plan between 1980 and 1985.

The need for efficient farming has to be balanced against the need for employment and a satisfactory quality of life in the countryside. In 1985 the agricultural sector remained the kingdom's largest employer, but it provided only about 3 per cent of the gross domestic product. Most of the people involved in farming were also unskilled, semi-nomadic and often illiterate. They would find difficulty in getting other work if new methods of farming were introduced, and would probably drift to the cities in search of employment. Economic development must be undertaken with these social consequences in mind.

The importance of irrigation

In the mid-1970s there were 150 000 hectares of land under crops, about a quarter of which was irrigated. During the Third Development Plan billions of dollars were spent to increase this, and there are now over 2 million hectares of cultivated land, most watered by modern irrigation methods.

The most spectacular change has been in the growing of wheat. Until the end of 1984 the government bought wheat grown by Saudi farmers at a very high price, far more than what it cost to produce and many times more than imported wheat. The incentive to cultivate wheat was therefore very great, and with money available the latest agricultural technology could be used. The result was enormous growth in production, so much so that some grain could be exported to neighbouring countries and some given as aid to

A livestock market in Riyadh

African states suffering from drought. Wheat subsidies have now been lowered, and farmers are being encouraged to grow other crops, such as barley for fodder.

Water for irrigation has mostly been taken from aquifers, or underground sources, near the traditional crop growing areas and in an arc from near the northern frontier with Jordan, through the area around Riyadh and into the south-west near the Yemen border. Most new programmes use the 'central pivot' method whereby irrigation arms up to half a kilometre long water a huge circular patch of land. There are over 15 000 of these in Saudi Arabia, and viewed from the air they provide spectacular patterns of green cropland in the otherwise brown desert landscapes.

There are concerns that these water resources might be over-used. Already wells have to be sunk to depths of 1500 metres to overcome salinity and poor water pressure. Overpumping in wells nearer the coast has also allowed sea water to seep into the underground aquifers, spoiling the supply. New wells can only be drilled under licence from the government.

Other agricultural developments

With financial help from the government farmers have been encouraged to buy land and buildings, machinery and irrigation pumps, and to get proper veterinary help with their animals. They are also able to use high quality seeds, fertilizers and insecticides, like the agribusinesses of the developed world. As a result of all this investment and change there has been a big increase in vegetable growing, poultry farming and dairying. The country is virtually self-sufficient in eggs, poultry and milk.

Much Saudi farming is now in the hands of large development companies. These involve foreign investment as well as local Saudi money, and sometimes employ farm managers with experience of Western farming methods. A lot of the farm labour is from the Philippines and other countries of the Far East. Future agricultural development is likely to concentrate on horticulture and the integration of fodder growing and

Modern farming using irrigation on the outskirts of Riyadh

livestock farming. None of these agricultural developments would have been possible without the discovery, exploitation and sale world-wide of another vital underground resource – oil.

1 Why do many countries attempt to be self-sufficient in food production when cheaper food is often available from elsewhere?
2 Describe the landscape and farming scene in the photograph. What are the signs that it is a recent development?

This graph shows the spectacular increase in Saudi wheat production

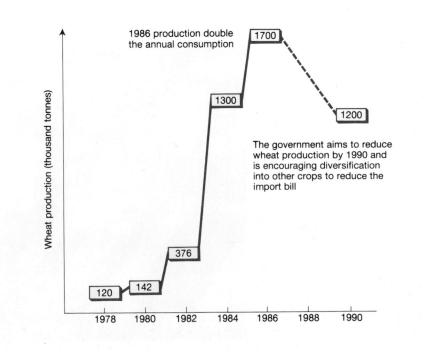

1986 production double the annual consumption

The government aims to reduce wheat production by 1990 and is encouraging diversification into other crops to reduce the import bill

151

3 Oil

Arab wealth

Outside the Riyadh palace of Prince Sultan bin Salman bin Abdul Aziz are the symbols of new Arab wealth: three Mercedes 500 SELs, a BMW and half a dozen American limousines. Inside the recently built palace, freshly decorated at a cost of $15m, are marble floors, a swimming pool, silk carpets, glittering chandeliers and deep cushioned sofas.

The 5 000 princes in the royal house of Saud have done well out of the oil boom; but then so have ordinary Saudis.

The Saudi Arabian government is a major arms purchaser

In the late 1930s, just before the outbreak of the Second World War, oil was discovered under the sands of the eastern deserts of Saudi Arabia. Within a few years Saudi Arabia had become an important exporter of oil to the industrialised nations of Europe and North America. The pastoral economy became transformed as the country suddenly became extremely rich.

The fortunes of all the oil exporting countries, especially those in the group known as OPEC, took an even more spectacular course following the Arab–Israeli War of 1973. The OPEC countries realised that they had virtual control over the resource that could make or destroy the economies of the industrialised world. The price of oil was greatly increased, quadrupling from $3 to over $11 a barrel in the space of three months in that year. Subsequently it rose much higher. The increase in income was enormous. In 1972 the total income from the sale of oil by all the North African and Middle East countries, from Algeria to Iran, was $10.5 billion. By 1981 this had increased to an annual income of $225 billion.

The waterfront at Jeddah

The impact of increased wealth

An immediate effect of these huge increases in oil prices was that the leaders of the country became enormously wealthy. The extract gives an indication of the personal fortunes amassed by members of the royal household and many others. But it also points out that most of the population benefited as well.

The vital contribution of wealth from oil to developments in agriculture and the efficient use of water resources has already been mentioned. Fresh water has been made available to the growing towns and cities through the use of expensive desalination plant that removes the salt from sea water. There has been a boom in building, frequently using foreign labour for construction work. Apart from houses and high-rise flats there have been developments of commercial and industrial buildings, roads, railways, ports and airports, and power stations.

There were great improvements in the provision of free education and health services. This has led to a large drop in illiteracy and a big increase in college enrolments. It is estimated that the number of doctors and hospitals per

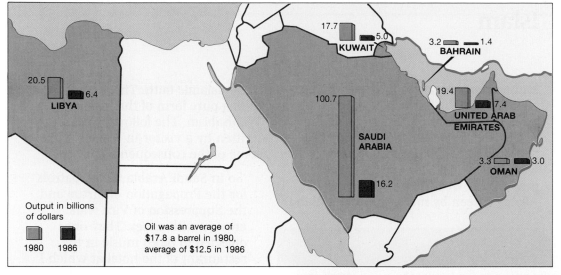

The decline in Middle Eastern oil earnings during the 1980s

KUWAIT 17.7 5.0

BAHRAIN 3.2 1.4

LIBYA 20.5 6.4

SAUDI ARABIA 100.7 16.2

UNITED ARAB EMIRATES 19.4 7.4

OMAN 3.3 3.0

Output in billions of dollars
1980 1986

Oil was an average of $17.8 a barrel in 1980, average of $12.5 in 1986

capita doubled as a result of the new wealth. Like most countries Saudi Arabia has spent a great deal on purchasing weapons, and has been able to buy some of the most effective and expensive aircraft, tanks and rockets available.

The end of the boom?

By 1986 it began to look as though oil wealth might disappear as quickly as it had appeared. Oil prices that had been over $40 a barrel slipped to below $10 a barrel as the producing countries struggled to retain their markets and undercut one another. World demand for oil had dropped, and there was too much oil being produced. Like other oil exporters, Saudi Arabia hoped to keep its markets by reducing prices. With a quarter of the known world oil reserves beneath its surface, Saudi Arabia can afford to wait for things to change. In 1981 its income from oil was $108 billion, but by 1985 this had fallen to $22 billion. Its financial reserves were also being used up at a remarkable rate, in spite of cutting its spending on construction and social programmes.

From everywhere in the Gulf there were reports of projects being postponed, of roads and railways not being built. Foreign workers were having to return to their own lands. For the very rich the impact was not too great, but overall the effect was beginning to be felt as the boom years of the late 1970s and early 1980s were left behind. As Prince Fahad

The price of oil may rise and fall, but Saudi Arabia remains a consumer society as these hoardings on the road from Jeddah to Riyadh indicate

bin Khaled bin Abdullah, himself a millionaire at 30 years old, said 'Our people must realise what happened is a dream which happens rarely in history. That dream is over and we have to adjust to reality.'

1 a) Graph the figures given on the map, and present them on a statistical map. b) Why did Saudi Arabia wait a long time before engaging in an oil price 'war' with other OPEC countries?
2 Who decides on how national wealth is spent in a) Saudi Arabia and b) the United Kingdom? What are the similarities and differences between the methods used in the two countries?

4 Islam

Muhammad, regarded as the messenger of Allah and the founder of Islam, was born in Mecca and later fled to Medina. These two holy cities of Islam are in Saudi Arabia. It is hardly surprising that with this historic link, and with its present day wealth and power, Saudi Arabia is seen by many as the heartland of the Islamic faith. The country practices a pure form of the faith known as Wahhabism. The following extract, written by a visitor in 1986, details some of the consequences.

'So in Saudi Arabia today, patrols for the Propagation of Virtue and the Suppression of Vice walk around with canes. They once objected to piped music in the restaurant of the hotel at which I stayed. The occasional thief has a hand chopped off. The occasional foreigner is flogged for making alcohol. The censors have a fine time. At a hotel bookstall I found a copy of *Private Eye* where scandalous references to Mrs Thatcher had been allowed to stand, and where Joanna Lumley's bare shoulders had been left bare, but where some Wahhabi hand had with a black felt-tip pen blotted out the decolletage of the Princess of Wales.

Women cannot work alongside men. They can on no account drive. Almost all go veiled with a black cloth which altogether covers the whole face and head. They cannot eat in public with men ... and although women can work with other women in schools and hospitals, they have to offer their services discreetly.

But the veiling of women, alcohol, the marks of felt-tip pens, and the odd hand cut off are trivialities. What is not trivial is that Saudi Arabia profoundly believes that its destiny is to lead the Muslim World.'

The Guardian. 1st April 1986.

Outdoor prayer at a camel race, Riyadh. Daily worship is an important feature of Saudi life

The five pillars of Islam

Every Muslim is guided in his religious and social life by the five pillars, the anchor points of Islam.

1. **To bear witness** that there is no God but Allah, and Muhammad is His messenger. This has to be recited sincerely and aloud at least once in a lifetime.
2. **To pray five times a day.** Whether a Muslim is praying in a mosque, in the home or at his workplace, he is required to face in the direction of Mecca. Friday is reserved for special congregational prayers in a mosque.
3. **To fast** during the month of Ramadan. Muslims observe a total fast from dawn to dusk for one lunar month (29 to 30 days) as a means of self-purification and an annual cleaning of the digestive system.
4. **To pay a divine tax,** called Zakat, for the 'purification' of their worldly wealth. The money is used for alms and other community needs.
5. **To perform the pilgrimage** – the *hajj* – to Mecca at least once in a lifetime by those who can afford it.

The place of women

It is generally felt in Western countries that women are unfairly treated and discriminated against in Islamic societies. The compulsory wearing of veils, the denial of basic education to girls and the general segregation of

women from daily activity is regarded as degrading. But it has to be acknowledged that many women – although by no means all – have welcomed a stricter reinforcement of such behaviour. Some of these rules are not a part of the original beliefs of Islam. The Koran, the holy book of Islam, in fact represented a big advance for women who had been regarded up till then as male possessions. Muhammad gave women a clear legal status as independent human beings, with strict rules about marriage contracts and the protection of dowries in the case of widowhood or divorce. Polygamy was permitted only if all wives were treated equally and justly. Women were allowed to control their own money without reference to husband or father long before this was the case in Britain.

Some countries, such as Egypt, decided that women should enjoy equal rights with men, and take a full share in the life of their country. Others have been more conservative, and in some cases have returned to a more fundamental faith that opposes women taking an equal role in society.

Saudi Arabia, the heartland of Islam, is one of the more conservative countries in its attitude towards the position and permitted behaviour of women. The Sari'ah law is applied in all its vigour. Most women live in seclusion, almost entirely within the family. But King Faisal's wife and daughters took the lead in opening the country's first private school for girls in 1955. By the early 1980s women were beginning to take

up more responsible positions, but Saudi Arabia still has one of the lowest ratios of working women anywhere in the world. There are now openings for jobs as teachers, social workers, doctors, and so on, but women still normally work in all-female institutions.

The place of women in Islamic societies raises difficult questions about development and its meaning. Saudi Arabia has acquired great wealth through the sale of oil, and many Saudis are benefiting in many ways. Yet in terms of democracy, with people electing their own government, and the rights of women, the country falls a long way behind many others in the world. But who has the right to say what should constitute social development?

1 What are the advantages to **a**) a country and **b**) the women of a country of having true equality of opportunities for all?
2 Do you think any group of people have the right to criticise others who hold different beliefs and behave in different ways?

(Above left) afternoon tea at the oasis is a men-only affair. (Above) jobs for women are often only available where they deal in services for other women, as at this 'Women's Branch' of the National Commercial Bank

The King of Saudi Arabia, one of the wealthiest men in the world, at a race meeting. Notice how closely guarded he is

ETHIOPIA
1 Environment and history

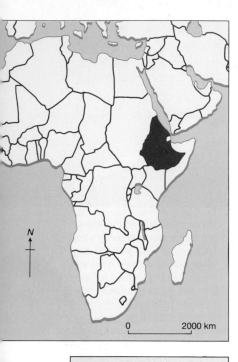

For large-scale map, see page 120

A rural landscape in Ethiopia. Cultivation takes place on the plateaux and along the floor of the river valley

The Republic of Ethiopia occupies much of the eastern part of northern Africa known as the 'Horn of Africa'. The northern province of Eritrea has a coastline along the Red Sea. Neighbours to the east are Djibouti and Somalia, to the south Kenya, and to the west Sudan. In 1986 its population was estimated to be about 42.3 million, the great majority living in scattered farms and small villages in the countryside. The largest of the cities is the capital, Addis Ababa.

Physical environment

Much of the country consists of large mountain ranges and high plateaux. The central plateau areas are at altitudes of between 1500 and 2000 metres (4900 and 6500 feet) and are cut into by deep river valleys. Ethiopia's highest peak is Ras Dashan, and at 4620 metres (15 154 feet) it is well over three times as high as Britain's highest mountain. These central highlands and mountains are usually rainy. The natural vegetation cover is forest, but much has been cut down to provide farmland and fuelwood.

The land falls to foothills and a narrow coastal plain along the Red Sea coastline, and to very dry foothills and semi-desert in the Ogaden region that adjoins Somalia. In these foothills the forest is replaced by more open woodland and scrub.

Perhaps the most notorious feature of the Ethiopian environment is the unreliability of the rains. Sometimes this can be very heavy and lead to erosion of the over-cropped and over-grazed hillsides, but in recent decades it has been the droughts of 1974 and the mid-1980s that have led to the greatest environmental and human disasters.

History

Ethiopia is one of the oldest of Africa's states, and has been ruled as a kingdom for several thousand years. Christianity became a major religious influence in about the sixth century AD, and the Christian Orthodox Church remained powerful and a large landowner until very recently. There is a larger proportion of Muslims in the areas away from the central highlands and the capital. They make up about half of the population of Eritrea, for example.

During the second half of the last century the major European powers forcibly took control of various parts of Africa. After the opening of the Suez Canal in 1869 control of the Red Sea became important, and Italy tried to gain a foothold on the continent in that area. Eritrea became an Italian colony in 1889, but early attempts to invade the kingdom of Abyssinia, as Ethiopia was then known, failed. The Italians finally took Abyssinia by force in 1935. The Emperor Haile Selassie was forced into exile and the three territories of Abyssinia, Eritrea and Somalia were incorporated into Italian East Africa. Italian rule was not to last for long, and in 1941, during the Second World War, Britain took over Eritrea and Emperor Haile Selassie was reinstated in Abyssinia. In 1952, with the agreement of the United Nations, Eritrea was incorporated into the state of Ethiopia and ruled from Addis Ababa.

The Emperor ruled over his kingdom in such a way that the people of Ethiopia seemed little affected by the changes taking place elsewhere in Africa. Many felt it was an unjust society, and in 1974 a violent revolution took place and thousands of people were executed, exiled or put in prison. The country since then has been ruled by a Head of State and military council that wants to introduce a communist way of life. As a result there has been some reluctance by Western nations to give aid, although on several occasions it has been desperately needed. On the other hand the government does get aid from the USSR, China and Cuba. Apart from the problems of coping with environmental and social problems, there has been a war with Somalia over territory in the east, and armed struggle against rebels demanding greater independence for Eritrea and Tigre.

Ethiopia, for these various reasons, is one of Africa's poorest countries. The great majority still try to live from farming, but the country is not self-sufficient in providing food. One of the few exports is coffee, but this brings in only a small income. There is little manufacturing industry. And in spite of the international airport at Addis Ababa and recent attempts to improve road and rail links, communications are still very poor for a country wanting to develop economically and socially.

Gondar market

Seventh century art in a church in Gondar. Christianity has been the major religion of Ethiopia for 1500 years

1 List some of the things that make development difficult for Ethiopia under the headings of **a**) the environment and **b**) the people.
2 How might being a communist state encourage or discourage aid being provided by the outside world?

157

2 Drought and famine

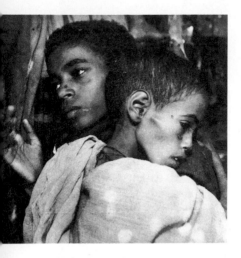

The face of starvation

The Ethiopians living in the highlands of Wollo and the neighbouring provinces have traditionally been reluctant to move from their scattered villages, but in January 1983 many heads of families were to be seen leaving the area seeking work or some means of survival for their families. It soon became clear that a disaster was imminent. After a short while whole families began to arrive at food distribution areas, some having walked between 50 and 100 miles. Within one week about 10 000 gathered in the town of Kobo, swelling the population to double its normal size. Large groups of starving people gathered at other points along the roadside, and it became clear that the disaster was becoming as serious as the previous one ten years earlier. During that famine over 200 000 Ethiopians had died.

Governments and voluntary aid agencies from around the world contributed help in the form of money, blankets, tents, fuel, cooking pots, light vehicles, medical supplies and food. There was a great surge of public concern, culminating in spectacular fundraising efforts such as 'Band Aid'. In the meantime hundreds of field workers continued their practical support away from the glare of publicity in the camps and villages in the famine areas.

A tent camp for victims of the famine

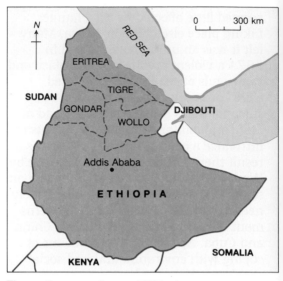

The northern provinces of Ethiopia

The scale of the drought and famine grew during 1984 and 1985. There were many deaths, and tens of thousands of starving people from the northern provinces of Tigre, Wollo, and Gondar sought help at Ethiopian government distribution centres. They were places where food and a little basic medical care could be obtained. One of the most spectacular of these tent cities was Korem in Wollo province. At the height of the famine there were over 100 000 people gathered at Korem, and they were completely dependent on aid from others. Korem is on the main road running from Addis Ababa northwards to Tigre and Eritrea, so aid supplies could reach the camp with comparative ease. Many villages and small towns were in much more remote areas, often perched on isolated plateaux. They could only be reached with great difficulty by road, and emergency supplies had to be flown in by helicopters or light aircraft, or dropped from low-flying transport planes. For many years there had been armed rebellion in these northern provinces, and anyone suspected of supporting the rebel forces was not helped. Their only hope of survival was to migrate into the Sudan, and many thousands of starving people walked across the frontier to the relative security of refugee camps in that country. The various forms of aid, from government sources and voluntary agencies, undoubtedly helped save the lives of hundreds of thousands of people, but the basic causes of the famine were not removed.

The causes of the famine

The immediate explanation of the famine was the failure of the rains, and the consequent inability to grow food crops or graze animals. Land in the provinces of Tigre, Wollo and Gondar in particular had been over-farmed, over-grazed and deforested. Much soil had been lost due to erosion of the overworked and exhausted land, and some hillsides that had once been productive were reduced to barren rock and scree. The villagers had to struggle to survive, and the failure of the rains was the last straw. Even if the rainfall proves reliable and adequate in the future it will be very difficult for people in these areas to build up the quality of the land and its resources to support a growing population at a higher standard of living.

One answer from the government is to resettle about 1.5 million people, taking them from the most eroded highlands to the relatively under-populated but malaria infested lowlands further south. At the same time it plans to plant trees and begin programmes of soil conservation and water management in the highlands. Many villagers do not want to leave their traditional environments and social groups for

Many people in Tigre province fled to neighbouring Sudan to escape both the famine and the fighting

something new and different. There is also great suspicion that the government is keen to move people from areas where rebellion is likely into large villages some distance away where their behaviour can be more closely observed and controlled. Most people accept that drought by itself was not the cause of the famine, but that the disaster had other economic and political explanations. A secure future for the area means these also have to be remedied.

An RAF plane on a 'mercy flight' dropping supplies

1   Imagine your family lived in a remote village in the highlands affected by the drought. What do you think your attitude to being resettled in a new large village a long way away might be?

3 Internal conflict

Eritrea and Tigre are the two most northerly provinces in Ethiopia. They were both very badly affected by the drought and famine of the mid-1980s. They are also both deeply involved in a violent struggle against the armed forces of the Ethiopian government. Civil war had been waged in Eritrea for almost twenty-five years, and in Tigre for almost ten, when the 1980s disaster added to their problems.

At the time of the drought most people in the two provinces lived in rural areas outside the towns. These were mostly under the control of the rebel forces and could not be governed or helped by the central government.

Independence for Eritrea

Eritrea was an Italian colony long before Italy conquered Abyssinia and created Italian East Africa. As the letter suggests, it was against the wishes of the Eritreans that they were incorporated into a new and enlarged Ethiopia in 1952. The Eritrean People's Liberation Front (EPLF) was set up to fight the armed forces of Ethiopia and obtain independence. It also has the aim of changing the nature of society so that there is less poverty and inequality and a greater stress on equal rights for everyone. Land reform is an important feature of these changes, and 'People's Assemblies' in the areas

EPLF troops in the frontline

controlled by the EPLF administer the programme. Many families that were originally landless now own their land.

The Eritrean Relief Association (ERA) was formed in 1976. It is the only voluntary agency working in the 'liberated' areas of Eritrea. It is represented on village communities that deal with matters such as farming, youth and women's affairs. The agency helps with water supply, vegetable gardening and animal husbandry and provides drugs and medical equipment. Organisations such as Christian Aid in the United Kingdom contribute to the work of the ERA, since it is politically difficult for governments to give aid to these 'rebel' forces fighting against the Ethiopian government.

The province of Tigre

Whereas the people of Eritrea want full independence, the Tigrean People's Liberation Front (TPLF) wants only to change the constitution of Ethiopia so that all the different nationalities within the country have a greater say in how it is run. They particularly want a greater say in the affairs of the province of Tigre, and are fighting the Ethiopian armed forces to achieve this.

The central highlands of Tigre are craggy, densely populated and very

At this clinic in Port Sudan, Eritrean victims of the conflict learn new skills

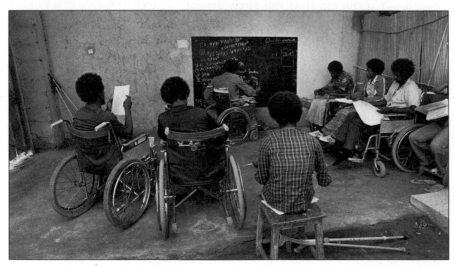

Eritrean struggle

Sir, I would like to draw attention to the plight of the people of Eritrea who are still carrying on their struggle for independence from Ethiopia. Eritrea was incorporated with Ethiopia – against the wishes of its people – as a result of the action of the United Nations.

In consequence of this and oppression by Ethiopa, in recent years supported by Cuban troops and other Soviet bloc 'military advisers', more than 150 000 Eritreans are estimated to have been slaughtered in the conflict and a further 500 000 driven into exile.

The time has come for the problem to be tackled at its roots by recognising the rights to self-determination of those who, without consultation and against their wishes, were forcibly included within the Ethiopian Empire – which the Soviet Union is using as the principal base for its military and strategic domination of the Horn of Africa, regardless of the cost in terms of loss of life and human misery.

From a letter by Winston S. Churchill, MP (Con).

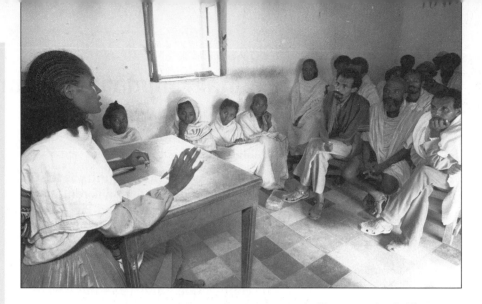

A village meeting in Tigre to discuss the effect of the drought on local education

badly eroded in many parts. As in Eritrea, much of the countryside is not under the control of the Ethiopian government. The TPLF has established a local administrative structure throughout the rural areas, and village councils have been set up to encourage local people to share in making decisions. The Relief Society of Tigre, which is similar to the ERA in Eritrea, was set up in 1978. It raises funds outside the country, and through a few full-time co-ordinators and hundreds of volunteer workers helps with health care, education, farming, forestry and terracing projects, and with resettlement schemes.

Farming provides the livelihood of over 90 per cent of the people of Tigre. Most grow little more than the food needed to survive even in normal times. The Ethiopian government has moved many Tigreans from the areas they do control and resettled them in other parts of the country. The reason is said to be to give the villagers a better chance away from the over-populated areas, but many feel it is to enable the people who might support the rebel cause to be removed and more carefully controlled elsewhere.

So many Tigrean men have joined the TPLF army, or been killed in the fighting, or have fled to Sudan, that the work and responsibilities of women have changed a great deal. They still have to do the back-breaking jobs of labouring in the fields, carrying water, and preparing food and running the family home. But they also now hold more responsible positions, filling 30 per cent of elected positions on the 'People's Councils' and taking important jobs in education and as lawyers. The people of Eritrea and Tigre, and the rest of Ethiopia, are spending resources and energy and human lives in the civil wars, but some economic and social benefits are to be seen even while the conflict continues.

1 What is the view of the letter writer about the civil war in Eritrea? What do you think a letter from a member of the government of the USSR might say? Why are they likely to differ?
2 Name other groups fighting for independence or a greater say in their own affairs. Can fighting be always justified to gain their ends?

A new road being built in Tigre under TPLF supervision

4 Feudal kingdom to communist state

'Villagisation'

In the south-east of the country is the province of Hararge, very different in many ways from Eritrea and Tigre. It covers about a quarter of Ethiopia and has a population of about 3.5 million. There are a few towns, such as Harer and Dire Dawa, but most people live in the countryside. Until 1984 there were about 3000 villages in the province. They usually consisted of homesteads scattered amongst their own fields within about eight kilometres of the village centre. A homestead consisted of several huts, one for the family and others for animals and food storage. Farmers normally grew enough food for their families, although larger farms sold any surplus grain or vegetables in the nearest town. Except in years of severe drought the area was self-sufficient in food.

One problem of these scattered homesteads and villages was that families were isolated, schools were a long walking distance away, and there was rarely any health service available if someone fell ill. The government claimed that these services would be much easier to provide if all the homesteads were clustered together, and so began the

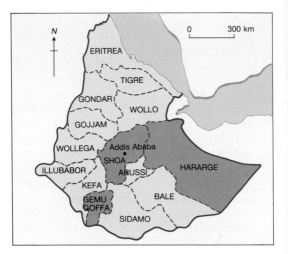

The provinces of Ethiopia, showing Hararge

process known as villagisation. During 1985 alone 3 million people were moved, often against their will, into new villages. Under the direction of local members of the Ethiopian Workers Party the villagers had to build their new homes in a central area, and then top them with the roofs from their old homestead. The houses are close together and usually without shelter for animals and food storage. All this work had to be done while still tending to the crops being planted, weeded and harvested back in the original village lands. Many families have to grow food on their old lands, and spend many hours walking to and from the new village each day, wasting much time and energy. Meanwhile the promised education and health services have had to wait.

The circumstances of villagisation, and the consequent change in the landscape and way of life of villagers, differ from those of the resettlement programmes further north in Tigre and neighbouring provinces. There is no problem of famine. It is instead a result of government policy to encourage or force people to live in these clustered new villages. A similar process has started in the southern province of Gemu Goffa and in the central province of Shoa. Over 5 million people may be moved in these two provinces alone.

A new village in the Hararge Highlands, constructed under the 'villagisation' scheme

There has been criticism of and resistance to the villagisation policy. The Ministry of Agriculture claims that while the policy was sensible, it was their responsibility and not that of local party members. The moves have been rushed, causing great loss of food production and resentment from the villagers. Those villagers who resisted were intimidated by such methods as having their homestead roofs set on fire. Others who refused to co-operate fled into nearby Somalia as refugees. It is claimed by some that the motive of the government was not to provide better services and improved quality of life, but to obtain greater control of the people through local Workers Party members in the clustered villages.

A communist state

Religion has had a strong influence in this part of Africa for many centuries, and the Ethiopian Orthodox Christian Church still has an important place in the state. But at the same time, in Revolution Square in Addis Ababa, portraits of Karl Marx, Engels and Lenin, leaders of the Russian Revolution, look down from bright red hoardings. The leaders of the military government are certainly backed by the USSR, and there are many Soviet technicians, advisers and politicians in the country. Ethiopia also gets support from Western governments as well as loans from the World Bank. Hundreds of Ethiopian students are sent to Western as well as Soviet universities. The revolution that overthrew the feudal type of government of Emperor Haile Selassie was not an uprising of the masses, but a take-over by young military officers wanting to create a socialist state.

It was not until 1984 that the military government set up the Ethiopian Workers Party to try and gain the wider support of people in the countryside and the towns. Villagisation is just one of the methods by which the government hopes to make a communist state in Africa. The general view is that there is less official corruption and inequality in Ethiopia than in many African states, but the price to be paid is greater state interference in everyday

A triumphal arch in the centre of Addis Ababa

life. The government looks powerful enough to ignore opponents of its policies. It is sometimes argued that the drought and famine which led to a great deal of Western aid not only saved millions of lives, but the military government as well. The USSR are keen to see the experiment succeed so that it can be an example to the rest of Africa of the benefits of a communist government.

1 Give examples of other planned movements to new villages and towns. Compare the methods and motives with Ethiopian villagisation.
2 Explain the message of the cartoon. Is it right or wrong for other countries to become involved in Ethiopia's affairs? Say why.

This cartoon appeared in *The Times*, and illustrates what many people believe happens to aid aimed at famine relief

NIGERIA
1 Environment and history

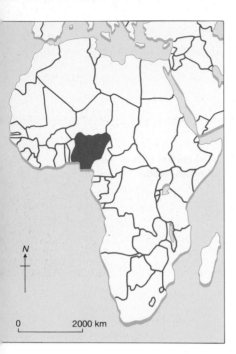

The Federal Republic of Nigeria is a very large country, almost four times the area of the United Kingdom, in West Africa. Its neighbours are Niger and Chad to the north, Benin in the west and Cameroon in the east. It is divided into nineteen states, plus the federal capital territory of Abuja.

The people

It is the most populous country in Africa, although an accurate figure is not known. The figures given by the censuses of 1962 and 1973 have been disputed. In 1983 the United Nations gave an estimate of 82 million, while the International Monetary Fund in the same year estimated a total of 89 million! It is now thought to be about 100 million. It is often not realised in Western nations how difficult it is to carry out a reliable census in most Third World countries. In Nigeria there are considerable differences between the peoples of the different parts of the country. Islam is the dominant religion of the north, and even then there are differences between, say, the Hausa and Fulani people. In the south the main groups are the Yoruba and Ibo, and both Christianity and local religions are more significant than Islam. Apart from these there are many other distinct groups within Nigeria with their own cultures, languages and ways of life. The official language of the country is English, a reflection of the history of West Africa.

Contrasts in environment

Nigeria's long coastline consists mostly of sand dunes backed by lagoons and a gently rising coastal plain. Towards the east is the large delta of the River Niger, with its dozens of distributaries threading their way through mangrove swamplands. The whole coastal area was once covered with rainforest, and many parts remain uncleared. The coastlands are hot and humid with an annual rainfall of well over 3000 millimetres (120 inches). This falls throughout the year although it is heaviest between June and October.

Inland the coastal plain is replaced by a higher and more hilly landscape. Along the border with Cameroon this becomes a large mountain range with peaks rising to over 2000 metres. Moving northwards the vegetation that remains uncleared for farming changes from dense rainforest to more open woodland known as savanna. The total rainfall is less than further south, and there is a marked drier period between November and March. The great River Niger flows across this central zone from

A market scene in the south of Nigeria

Contrasting environments in Nigeria. (Above left) the rainforest of the south-eastern coastlands. (Above) farmland on the dry savanna of the north

the north-west. It is joined by the major tributary, the River Benue, which flows from the north-east, just south of the large high upland area known as the Jos plateau. The combined waters then flow southwards and enter the Gulf of Guinea through the main channel and distributaries forming the delta. As can be seen from the photographs, the savanna landscape around Zaria, with its scattered trees, is very different from the forest lands nearer the coast.

Further north still the land falls to the low plains of Niger and Chad. The rainfall here is very much lower, often totalling less than 500 millimetres (20 inches) in a year. This area borders the Sahel, with its notoriously unreliable rainfall and long period without precipitation between about November and March. The vegetation gets correspondingly sparse, and is little more than desert scrub in the far north during the dry season or failure of the rains.

The Federal Republic

The main reason why such diverse people living in such contrasting environments should belong to one state is largely due to the European colonisation of West Africa. European traders visited the coastlands from at least the fifteenth century, seeking gold and slaves. With the decline of slaving, trade in forest products such as palm-oil became more important. In 1861 Lagos Island became a British colony and within a few years Britain had acquired, through treaties with local chiefs, almost exclusive trading rights along the coast. First the southern areas and then the northern territories became British protectorates. In 1914 these separate provinces were combined as the colony of Nigeria, although continuing to be administered in different ways. Nigeria, along with other European colonies, sought independence after the Second World War, and became self-governing in 1960. There has been suspicion between the different groups, and even civil war, since independence. There has also been a succession of military and civil governments, the ninth being a military one taking power in 1984. One of the most important needs of any government is to weld the diverse people of this artificial and externally mapped country into a unified whole, and be seen not to favour any particular tribe, religion or state.

1 In what sense is Nigeria an 'artificial' state? Give other examples of states with people of different cultures, religions and languages.
2 What makes a 'natural' state? What is a 'good' boundary? Is it helpful or not to have a variety of environments and peoples within a state?

For large-scale map, see page 121

2 The green revolution

Crisis in Nigerian agriculture

The wide variety of environments, ranging from equatorial rainforest to savanna and semi-desert, means that a great variety of crops can be grown in Nigeria. Many, such as yams, cassava and vegetables, are grown for the family or sale in local markets. Others are grown for export. Good examples of these are cotton and groundnuts from the drier north, and cocoa, rubber and palm-oil from the rainier south. These export or cash crops may be grown on small family farms or on larger plantations.

Modern machinery is most commonly found in the production of major cash crops, like the oil-palm (below), while in the north of the country cattle rearing remains an important activity (bottom)

It is thought that about 70 per cent of the Nigerian labour force works in agriculture, giving some idea of its importance. It is estimated that about three-quarters of this huge country could be cultivated, although only about a third is actually used for growing crops. Within these crop growing areas there is often great pressure on the land, with dangers of soil exhaustion and erosion. Cattle farming is found mainly in northern parts since the animals are affected by tsetse-fly in the more humid south.

Nigeria is more fortunate than many Third World Countries in terms of potential for agricultural development, but in recent decades things have not gone well. While the population has grown at a fast rate, there has been little increase in food production. Many explanations are offered. It is claimed that there is a shortage of farm labour as young people move to the cities in search of a more interesting and rewarding life. It is said that many farmers lack knowledge of new and better methods of farming, and are unable or unwilling to use new types of seeds and fertilizer. In many places there is need for irrigation, and a better road system to transport produce. Traditional patterns of land ownership, with small and scattered fields, makes the use of machinery more difficult. Rightly or wrongly it is argued that many farmers are both conservative and inefficient, and that what is needed is larger farm size, modern technology and modern equipment.

Quite apart from this, the high exchange rate of Nigerian money has meant a large increase in imports of cheap foods from abroad. The pattern of diet also changed during the years of oil wealth, with many Nigerians changing from traditional foods to rice and bread. At the same time as food production for local consumption has stagnated there has been a remarkable fall in exports of food crops, especially since the beginning of the 1980s. Once again there have been many reasons. Many of the plantations in the south were damaged during the civil war, and have never recovered. While oil exports were earning so much wealth the government was less concerned about supporting agricultural exports. As the price offered for the cocoa and palm-oil, groundnuts and rubber, cotton and export grains fell on world markets, the farmers were unable to obtain a worthwhile income for their efforts.

In 1985 imports of rice, maize and vegetable oil were banned to encourage higher local production. Rice production has risen rapidly.

The Green Revolution

The government has tried many ways to make Nigeria more self-sufficient in foodstuffs and cash crops. During the 1970s a programme using lots of extra temporary labour was tried, called 'Operation Feed the Nation'. Land Use Acts were passed to try and change the pattern of land ownership. State food-farming has been tried, several major and very expensive irrigation schemes have been built, and a number of 'Integrated Agricultural Development Projects' begun. These IADPs try to combine help to farmers through grants and loans with the building of rural roads and water supplies. None of these methods have been very successful. The irrigation schemes are very expensive to build and run, and do not always make it easier for the local farmers to improve their output. It is said that the IADPs only help the larger farmers while the poor peasant farmers fall even further behind.

The phrase 'green revolution' has been used to describe attempts worldwide to improve agricultural production through using high-yield crop varieties together with heavy use of fertilizer and pesticides. In Nigeria it is used to describe a change of emphasis from big projects to help for small farmers – who produce about 90 per cent of the output.

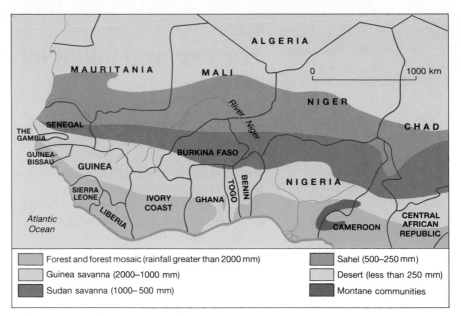

A vegetation map of West Africa

Through giving subsidies for the crops produced and encouraging the use of simple implements and animal power rather than relying on foreign expertise and complicated technology the hope is that millions of ordinary farmers will become more efficient and produce enough food for Nigeria to become self-sufficient.

1 What is meant by 'appropriate technology' when applied to farming? What are the advantages to the small farmer of help in the form of appropriate technology compared with help from major projects such as a big irrigation scheme?
2 What might encourage farmers to grow more crops for export? In what ways might a big increase in export crop production hinder Nigeria in becoming self-sufficient in food production?

Pyramids of groundnut (peanut) sacks line the road in Kano, northern Nigeria

167

3 The rise and fall of oil wealth

Before the 1960s Nigeria relied heavily on farming for its wealth. Feeding the population was not a major problem, and food imports were restricted to a few luxury items. Farm produce also provided most of the export income. Cocoa, groundnuts and palm-oil products contributed almost three-quarters of export earnings, while the balance was made up by rubber, cotton and other agricultural produce. The average income per person was not very high, but the economy was based on a wide range of products and the country was fairly self-sufficient.

The commercial production and export of oil began in 1957, but it was during the five years between 1970 and 1975 that the economy was transformed. Due to decisions taken by the

OPEC countries the price of oil was raised very quickly in 1973. There was not only a four-fold increase in Nigerian production, but also a trebling in the income from oil as a result of the OPEC agreement. The pattern of Nigerian exports changed dramatically. Oil soon provided over 92 per cent of the export earnings, and the total amount earned increased enormously. Nigeria became very wealthy.

Many changes resulted from the vast amounts of money available to the federal and nineteen state governments. Salaries of civil servants were greatly increased, and a lot of money could be made from providing services to support the boom in trade. Because many of these activities are urban, there was a big movement of people from the countryside into the towns, as well as an increase in migrant labour from neighbouring countries.

A lot of the new wealth went into the building of ports and airports, bridges and roads, factories, and new education and health facilities. The

Oil tanks at Port Harcourt, with a tanker in the background

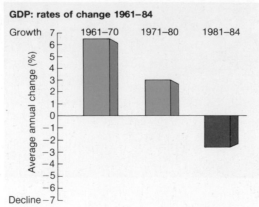

The fall in oil prices in the early 1980s severely affected the Nigerian Gross Domestic Product, as this graph shows

GDP: rates of change 1961–84

Nigerian oil production, 1963–82

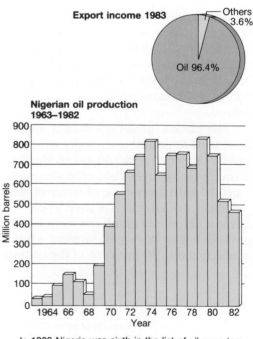

Export income 1983

Others 3.6%

Oil 96.4%

Nigerian oil production 1963–1982

In 1983 Nigeria was sixth in the list of oil exporters

exchange rate of the Nigerian Naira was kept high, which meant that imports were relatively cheap. As a result the factories tended to use imported raw materials rather than the more expensive supplies from within the country. In a similar way food imports increased, partly because they were fairly cheap, but also because the wealthier urban population began to prefer foods based on wheat, maize and rice, such as breakfast cereals, and tinned and processed foods.

The high exchange rates also meant that it became more expensive to buy Nigerian goods overseas, and so the export of groundnuts and palm-oil products fell very rapidly. Cotton was another export crop to decline during this time. In some cases the commercial farmers tried to sell their produce in local markets, but often they just ceased growing the crops on their farms and plantations. Generally speaking, as was seen in the previous section, agriculture stagnated while the rest of the economy grew with the boom in oil income.

The end of the oil boom

The dangers of dependence on one export began to be realised in the late 1970s. There was a world-wide decline in economic activity and world trade, and a decline in the demand for oil, partly caused by the leap in oil prices a few years earlier. The OPEC countries tried to share the loss of oil income by voluntary agreements on limiting production and keeping up prices, but Nigeria and a few others have been tempted to get what income they can by selling more cheaply than the OPEC figure. In spite of a brief upsurge in the early 1980s, the income from oil has continued to fall.

Expenditure did not fall at the same rate, and at first Nigeria managed by using its savings. When those revenues had been spent the country was forced to seek loans from commercial banks, the International Monetary Fund and the World Bank. Both military and civilian governments have tried to spend less on new building and industrial projects, and on schools, hospitals and social services. At the same time they

Investment is badly needed in Nigeria's major asset, its large population of young people

have tried to save by reducing imports through various controls. But this has forced factories dependent on imports to put up prices or close down. There has been a big increase in the cost of living and in unemployment. Major developments such as the large new iron and steelworks and railway link at Ajaokuta in Kwara State, and the new federal capital at Abuja, have been slowed down or postponed.

Various ways are being tried to halt this downward spiral into greater problems. Fresh loans have been arranged, plans to exchange oil for help with particular projects have been agreed with France, Italy, Brazil and Austria. But the proposal to lower the exchange rate has been resisted. Although it would immediately reduce the cost of imports it would lead to a big increase in the cost of living and make the government very unpopular. Nigeria is barely coping with the fall in the price of oil.

1 Explain the danger of depending on one major export commodity, using the example of oil in Nigeria.
2 Why do some people argue that the sudden income from oil was a bad thing for Nigeria?

The new federal capital at Abuja is being built from scratch, but has proved expensive and work has been suspended since the military took power

169

4 Links with the wider world

Economic trading patterns

An idea of the types of goods involved in Nigerian trade in the mid-1980s, and the main trading partners, can be gained from the tables. It is interesting to see how little trade there is with neighbouring African states – in fact it was only about 2 per cent of the total. This is a reflection of their similar environments, type of development, and colonial histories. Most West African countries are exporters of minerals and importers of manufactured or semi-processed goods.

The USA and several European countries are the major buyers of goods from Nigeria. What is striking about the major exporters to Nigeria is the importance of the United Kingdom and the inclusion of Japan. Trading patterns change from year to year, of course, and there was a marked decline in both exports and imports during the mid-1980s. Even so, international trade remains as vital to Nigeria as to most countries of the world.

Goods entering international trade have to be carried by road and rail, air, or by sea. Since so much Nigerian trade is with countries in other continents, sea and air transport dominate. Ship cargoes are handled at one of six major seaports – Apapa, Lagos Island, Port Harcourt, Warri, Sapele and Calabar. The two

Main trading partners, 1983			
Import sources	Per cent of total	Export destinations	Per cent of total
UK	21.0	USA	21.8
USA	14.2	France	20.7
W.Germany	13.3	Italy	13.8
France	12.5	W.Germany	13.3
Japan	9.3	Netherlands	8.9

main international airports are Ikeja, near Lagos, and Kano in the north of the country. There are many flights between the main towns of Nigeria, but internal transport from the interior to the ports is mainly by road.

Links with other West African countries

Nigeria is the most prosperous of the sixteen countries making up the Economic Community of West African States (ECOWAS). This association was founded in 1975 by the Treaty of Lagos, and the countries included are shown on the map. Nigeria is the dominant country, with 57 per cent of the population and 69 per cent of the GNP of the combined states. The administrative capital of ECOWAS is also in Nigeria, at Lagos.

The long-term aims of ECOWAS are to abolish trade restrictions between the member states, develop similar agricultural and monetary policies, and to share in research projects and in marketing. There is also the goal of free movement of labour, services and capital within the community. The difficulty of keeping to these aims is illustrated by Nigeria's decision to forcibly evict thousands of migrant workers from neighbouring states when unemployment rose due to the drop in oil income.

The modern city of Lagos is the major port through which goods enter and leave the country

Migrant workers from neighbouring states were forced to leave Nigeria in the mid-1980s when local unemployment levels rose

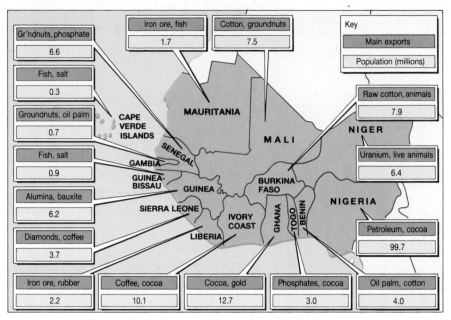

Country/region	Main exports	Population (millions)
	Gr'ndnuts, phosphate	6.6
	Fish, salt	0.3
	Groundnuts, oil palm	0.7
	Fish, salt	0.9
	Alumina, bauxite	6.2
	Diamonds, coffee	3.7
	Iron ore, fish	1.7
	Cotton, groundnuts	7.5
	Raw cotton, animals	7.9
	Uranium, live animals	6.4
	Petroleum, cocoa	99.7
	Iron ore, rubber	2.2
	Coffee, cocoa	10.1
	Cocoa, gold	12.7
	Phosphates, cocoa	3.0
	Oil palm, cotton	4.0

The major exports (and populations) of the ECOWAS countries

Co-operation is made difficult by the great differences in size, wealth, political systems and state of development of community members. The smaller and poorer countries inevitably depend upon the larger and richer. Nevertheless the wish exists to develop a community like that in Western Europe, and Nigeria is a key member.

International relations

Nigeria is a member of a number of other international organisations. Its membership of OPEC has already been mentioned. Some of the most influential members of OPEC are the Arab states in the Middle East. Another important group to which Nigeria belongs is the Organisation of African Unity (OAU), which attempts to bring member countries together to face mutual problems and share developments. It is also a member of the Commonwealth, the group of nations from all over the world that were once part of the British Empire and have voluntarily chosen to retain a co-operative relationship. Nigeria is one of the 65 African, Caribbean and Pacific signatories of the Lome Convention which seeks to develop better trade and aid relations with member states of the European Economic Community.

Most independent countries of the world are members of the United Nations. Although many of its efforts have proved disappointing, there are many hundreds of thousands of people who have benefited from its existence. Nigeria is a member of the UN, as well as the IMF and the World Bank. All these economic, cultural and political links are an indication that few countries can exist satisfactorily in isolation, and that there are usually many advantages to be gained from co-operation with others.

1 Which European countries were mostly responsible for creating states in West Africa? How does this help explain the nature of the Lome Convention? In which country was the convention signed?
2 What are some of the advantages and disadvantages of a) being involved in international trade and b) being a member of an international organisation?

Sporting links between African countries are important. In this Africa Cup match Nigeria were playing Cameroon

171

CUBA

1 Environment and history

For large-scale map, see page 122

The Republic of Cuba includes the large island of Cuba itself and over a thousand others ranging in size from the Isle of Youth to tiny coral reefs. Cuba lies in the northern part of the Caribbean Sea, only 144 kilometres (90 miles) off the southern tip of Florida, USA. This location has a powerful influence on Cuba's economy and development.

Physical environment

The main island is about 1200 kilometres long and 150 kilometres in width, although as the map shows this varies from place to place. The deeply-indented coastline consists of cliffs, bays and long strands of coral. Much of the interior is low-lying plain that was once covered with forest but is now mostly cultivated. These plains are interrupted by occasional uplands and mountain ridges. The main mountain ranges are the Sierra de Trinidad in the centre of the island, which rises to over 1000 metres (3300 feet), and the Sierra Maestra in the south-east. The highest peak is Pico Turquino in the Sierra Maestra range, which reaches an altitude of 2005 metres (6578 feet). Cuba has a sub-tropical climate, with the year-round high temperatures modified by sea breezes around the coasts and by high altitude in the mountains. It usually rains in every month, with a season of heavier rainfall between May and November.

History of the islands

When explorers from Spain first visited the islands in the late fifteenth century they were densely populated with local Indian people of different groups. They lived in small rural settlements and existed by hunting or collecting food from the forests. With superior force the Spaniards were able to take control of the islands, and they soon divided Cuba into large estates. The sugar plantations they created became a great source of wealth, but demanded a lot of manual labour. This could not be provided

Urban and rural Cuba. A street scene in Havana (below), and a view of the Sierra Maestra Mountains (below right)

locally, so large numbers of slaves were brought in from Africa.

Hatred of Spanish colonial rule increased during the following centuries, and led to the war of independence between 1895 and 1898. The USA was extremely interested in the future of this wealthy country just off its coast, and very soon after the Spaniards were defeated and relinquished control of the country, it was occupied by the USA. Cuba soon gained its independence, however, and in 1902 became a state with its own government. The USA retained naval bases on the island, and had a great influence on the economy, politics and way of life of the country. The hotels and clubs of Havana, the capital city, and the attractive beaches and resorts, became very popular with American tourists, for example.

A series of corrupt regimes ran the country during the first half of the twentieth century, and there was little economic or social development for the majority of the people. Between 1953 and 1959 a small group of Cuban nationalists, led by Fidel Castro and the Argentinian Che Guevara, conducted an armed insurrection against the government. The dictator Fulgencio Batista and his supporters were finally overthrown on New Year's Day 1959, and a Marxist government was established.

The country is divided into provinces and municipal areas. There are 169 municipal councils responsible for local affairs. It is calculated that at the last elections 98 per cent of the population voted. However, only the communist party is allowed to contest the 11 000 seats. Representatives from these councils are then elected onto provincial and national governments.

Population and people

The total population of Cuba in 1986 was estimated at 10.1 million, about 70 per cent living in urban areas. Havana has a population of almost 2 million, and Santiago de Cuba, Santa Clara and Camaguey each have around half a million. About 72 per cent of the people are white, mainly of Spanish origin. Of the remainder about 12 per cent are black, 15 per cent are mulattos of mixed

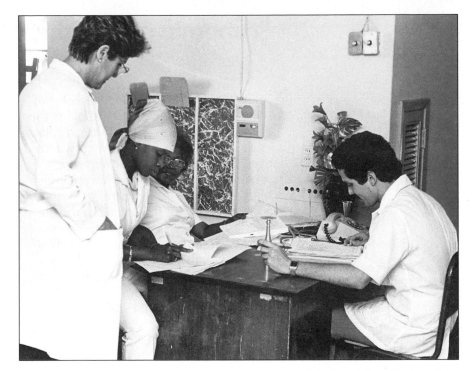

A nursing station at a hospital in the centre of Havana

European and Negro origin, and 1 per cent are Chinese or other ethnic groups. Although Cuba is a communist state it is thought that more than half the population practices some religious faith, with about 40 per cent being Catholics.

Castro's kingdom

In the 1950s Cuba was statistically one of the richer countries in Latin America. But away from Havana there was great poverty. Rural houses were palm-thatched huts lacking every basic service. It was to change this, and get rid of Batista, that Castro took the revolutionary path.

His first attempt was a disaster. In 1953, he led an attack on an army barracks in Santiago. It ended in a shambles with most of the participants killed. Castro was jailed for 15 years.

Released two years later he went into exile in Mexico. It was here that he met the Argentine Ernesto 'Che' Guevara.

His return to Cuba in 1956 was at first no more successful.

Missing the General Strike timed to coincide with their arrival, his troops landed in the wrong place and were decimated. Just a dozen men were left. Remarkably, just two years later Batista fled the country. The revolution had begun. For Castro consolidating it was going to prove even harder.

1 With the help of diagrams describe some of the characteristics of the Cuban population. Explain some of the ethnic and religious features.
2 Name other countries where an unpopular government or dictator has been overthrown following an armed uprising.

2 Economic development

In Cuba the government owns most of the means of production, and plans the economy. The only important exception in the mid-1980s was farming, where over 20 per cent of the farms were privately owned. This is a surprisingly high proportion in a centrally planned economy. As can be seen from the pie graph, about 21 per cent of the workforce is employed in farming, which is fairly low for a Third World country. This includes the 190 000 or so small-scale private farmers and those who work on collective and state farms.

These pie graphs showing Cuban employment and production figures present an interesting contrast

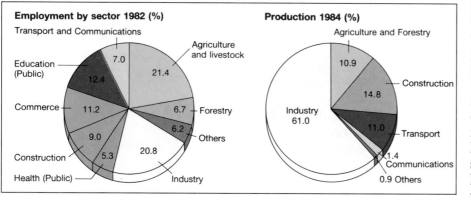

Employment by sector 1982 (%)

Transport and Communications 7.0
Agriculture and livestock 21.4
Education (Public) 12.4
Commerce 11.2
Forestry 6.7
Others 6.2
Construction 9.0
Health (Public) 5.3
Industry 20.8

Production 1984 (%)

Agriculture and Forestry 10.9
Construction 14.8
Industry 61.0
Transport 11.0
Communications 1.4
Others 0.9

Land reform

In pre-revolution times much of the farmland consisted of large commercial estates, known as latifundia. These were traditionally owned by wealthy families and worked in a semi-feudal fashion by impoverished farm workers. In later years they included large sugar-cane plantations and cattle ranches owned by Cuban or American companies, and understandably the profits went to the shareholders or were invested in the big estates. Only a third of the farms were worked by their owners. The remainder were farmed by tenants, share-croppers and landless farm workers.

Castro and his small invasion force promised land to the landless peasants, and it was their support that led to the success of the revolution. The first Agrarian Reform Law in 1959 limited the amount of land anyone could privately own. Excess land was distributed amongst the landless peasantry, while less efficient land became the property of the state. The overall reduction in farm size and the redistribution of land was impressive, and due to the reforms some 165 000 individually-owned farms were created covering nearly 50 per cent of the agricultural land. The rest was reorganised by the state. Sugar production was put in the hands of 600 co-operatives, and others were created for other crops and for mixed crop-livestock farming.

The second Agrarian Reform Law in 1963 eliminated medium-sized private farms by reducing the maximum amount of land a private farmer could own. The amount of land controlled by the state increased to about three-quarters of the total. Since then the state has acquired more private land on the

Cutting sugar cane (below). Cuba depends heavily on its sugar exports. But the climate is favourable for growing many other kinds of plant, such as the aloe (below right) which has medicinal uses

retirement or death of the owner, or by persuading the owner to join a co-operative farm. Private farmers are responsible for a large proportion of the tobacco, coffee, vegetable and beef output, so the government does not want to compulsorily take over their land. But there is little doubt that the complete collectivisation of agriculture is the ultimate aim.

Modern Cuba features newly constructed rural towns for workers on state farms. The plantations now provide higher wages and year-round employment, with improvements in housing, health and education services. Pupils in secondary schools devote half their time to lessons, and half to working the surrounding fields planted with citrus fruits, vegetables and other crops for the domestic market.

The importance of sugar-cane

Sugar-cane dominates the economy. Cuba is the second largest sugar producer after Brazil, and sugar provides about three-quarters of export earnings. About 85 per cent of the sugar-cane estates are owned by the state, and it is here that mechanisation of sowing, cutting, and lifting is most developed. In 1985 eleven new mills were built bringing the total to 163, and 40 of the older mills were re-equipped. During the same period two new refineries were opened, and the remaining sixteen modernised. Cuba is set a quota of sugar it can export, but in several years during the early 1980s it failed to reach its target. This was due partly to the very heavy rainfall during the harvesting season and partly to the inefficiency of the older machinery. Because of the need to earn foreign money some sugar is imported from other Caribbean countries to be refined and then re-exported up to the total that has been internationally agreed.

Mining and manufacturing

The pie graphs indicate the importance of mining and manufacturing to Cuba both in terms of employment and value of production. Cuba has the largest nickel reserves in the world, and is the

fifth largest producer. Production is mostly in the east with mines at Nicaro and Moa. A refinery was opened at Punta Gorda (Santiago de Cuba) at the end of 1985, and semi-processed nickel is shipped to the USSR. The USA trade embargo on Cuba prevents the import of any goods containing Cuban nickel. Cuba produces and exports other important minerals such as copper, chromite and cobalt, but is still not self-sufficient in iron ore or oil.

1 What seem to be some of the advantages and disadvantages of being a) a landless worker on a pre-revolution plantation, b) a small private farmer after the revolution, and c) a member of a modern collective or state farm?
2 What is the danger for Cuba of being so dependent on sugar for export income?

Newly-constructed state housing for workers on a state farm. Notice that this includes electricity supply

Havana harbour, with an oil refinery in the background

3 Trade and international relations

Imports (%)

Others 21.6

USSR 66.3

China 3.4
Czechoslovakia 2.4
Bulgaria 2.6
East Germany 3.7

Exports (%)

Others 13.8

USSR 72.1

China 3.7
Czechoslovakia 3.0
Bulgaria 3.4
East Germany 4.0

Sources of Cuban imports and destinations of Cuban exports

Consumer goods are in short supply in Cuba. This shop window displays imported second-hand Soviet TV parts

Trade

The bulk of Cuba's trade is with other socialist countries. Cuba belongs to the Council for Mutual Economic Assistance (CMEA), known in the West as COMECON. This group was formed in 1948, and nowadays consists of the USSR, Bulgaria, Czechoslovakia, East Germany, Hungary, Poland, Romania, Yugoslavia, Vietnam and Mongolia as well as Cuba. Afghanistan, Angola, Ethiopia, Laos, North Korea and South Yemen may attend meetings as observers. As the pie graphs show, Cuba's main trading partners in 1984 were members of CMEA. What is not shown are the invisible earnings Cuba receives from technical and military assistance given to some Third World socialist countries, particularly in Central and South America and Africa.

Sugar is by far the largest single export earner, contributing three-quarters of the total in 1984. The USSR is easily the biggest purchaser of Cuban sugar, as well as of nickel, an important element in steel-making. The USSR helps Cuba by paying prices above the rate on the international free market, as well as by selling oil and manufactured goods at relatively cheap prices. Much of Cuba's trade with the USSR and other CMEA countries is counter-trade, where goods are exchanged for other goods, rather than bought and sold using currency. Apart from granting these generous terms of trade, which protect Cuba from the big price fluctuations that often occur on world markets, the USSR also helps with military and other development aid. In the early 1960s Cuba was receiving some 60 per cent of the total aid given by the USSR to all countries. Cuba receives such support not only because it is a socialist country needing help to develop economically and socially, but also because the USSR considers it very useful to have friendly bases close to the USA, which it regards as its major threat.

The country's main exports are sugar, nickel, citrus fruits, tobacco and fish, and the main imports are machinery, transport equipment and fuels. In spite of exports being priced well above world averages, and imports often under-priced, Cuba has a history of trade deficit (spending more than it earns) with both socialist countries and those with market economies. Cuba has recently been forced to borrow from Western banks at whatever interest rates it can get. It has also tried to boost its trade with Western countries, and has sent and received trade delegations from many countries.

The government is keen to encourage foreign investment in most economic developments, and money from other countries is used for such things as offshore oil exploration and development, petrochemicals, steel manufacturing and tourism. Throughout the world tourism is nowadays big business, and Cuba has many sites capable of development. Centres for tourism are planned for many of the coral islands. Major new tourist centres are also being built with Mexican help at Cayo Largo and with French investment at a site about 100 kilometres east of Havana.

International relations

While there have been some disagreements with the USSR about the way socialism should be introduced, relations between the two countries have been

This hoarding in Havana says 'Mr Imperialists, we're not afraid of you at all'

friendly. Cuba might have found it difficult to survive without the considerable economic and military help from the Soviet Union. On the other hand the USA has always resented having a socialist country so friendly with the USSR so close to its coastline.

Many supporters of the overthrown Batista government sought asylum in America, and opponents of the revolution left freely by air until the United States stopped flights to Cuba in 1961. After that many left by sea during the 1960s and 1970s. In 1980 Castro decided to allow anyone who wanted to leave to do so. Some 125 000 people left, mainly on small boats from the port of Mariel west of Havana, for Florida. In the twenty-five or so years following the revolution around half a million people have left, the great majority settling in the United States.

Relations with the USA remain poor. In May 1985 radio broadcasts from Florida supporting opponents of the regime began, and in retaliation Cuba suspended an agreement made the

previous year which allowed the emigration of 20 000 Cubans a year and the repatriation of several thousand criminals and mentally-ill emigrants who had left in 1980. The USA has also imposed economic sanctions on Cuban trade, and refuses to import goods containing anything of Cuban origin. The suspicion and fear between these two countries does nothing to help the economic and social development of either.

Two Cubans in front of a squad of Angolan soldiers (some of them girls) they are helping to train

1 What are some of the ways in which development in Cuba has been greatly affected by the 'cold war' between the USSR and the USA?
2 What right has any country to interfere in the affairs of a neighbouring country, even if it has a different political regime that it dislikes?

Cuba is often described as being in 'America's backyard'. But these maps, on the same scale, show that the whole of Western Europe could equally be described as in 'Russia's backyard'

4 Social development

Measuring social development

Cuban economic wealth is not measured by gross national product as it is in most Western countries. Instead the Soviet indicators of gross material product (GMP) and gross social product (GSP) are used. Both include the value of agricultural and industrial production, while GSP also includes the value of material services such as transport, trade and communications. GSP, in spite of its name, excludes 'non-material services' such as education, health and defence, which are classified as consumption. The GSP of Cuba went up each year during the first half of the 1980s, although for the reasons just given, this does not necessarily mean there was an increase in living standards.

Social progress

The evidence of clinics, hospitals and schools suggest that Cuba has made exceptional progress in health care and education since the revolution. The provision of adequate housing for all has not been so successful, having been given a lower priority and handicapped by a shortage of resources. Unemployment is also low according to official figures, but with a growing population and a large number of people under 16 years of age the creation of jobs is likely to be a continuing problem. Quite a lot of surplus labour has been used by sending people abroad on military or development services, while quite a few dissidents who left in the 1980 emigration were workers who thought they could get better work and rewards in the United States.

Cubans measure their standards of living not against those of the Western consumer societies but against those of the rest of the Third World and their own pre-revolutionary past. In spite of difficulties such as the world-wide increase in oil prices, the fall in income from the export of sugar, the loss of sugar markets in the EEC (where sugar growers were being protected), and the

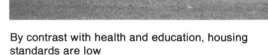
By contrast with health and education, housing standards are low

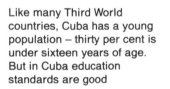
Like many Third World countries, Cuba has a young population – thirty per cent is under sixteen years of age. But in Cuba education standards are good

difficulty of getting short-term loans from Western banks, the general standard of living in Cuba has continued to improve.

A popular regime?

It is difficult for any outsider to know what it is really like to live in another country, particularly if it is organised according to different beliefs and a different political system. Some people argued that the large number who chose to leave Cuba in 1980 showed how dissatisfied people were. A different view is given in this extract written in 1983 by someone who already knew the country.

'Returning after an interval of four years one finds significant improvements in the supply of food and consumer goods. While the capitalist world rations goods by price and the Soviet Union by queues, Cuba uses the ration book. But in the last few years it has developed the so-called parallel market, under which people can buy goods above the ration at a considerably higher price. Some of the rations are generous, such as two pints of milk a day for every child under seven and six pounds of rice per person per month. Many products – eggs, bread, yoghurt and cream cheese – are not rationed.

These improvements take place against the background of Cuba's continuing commitment to a system of free education and medicine, which have given Cuba one of the lowest infant mortality rates and teacher/pupil ratios in Latin America.

A strong emphasis on egalitarianism marks Cuba out from most of its socialist allies. Raul Leon, President of Cuba's National Bank, said "We attach the greatest importance to the spirit of solidarity among our people ... It is part of our philosophy that we should not have substantial income differences."'

Jonathan Steele *The Guardian*
1st August 1983

While consumer goods such as electric fans, refrigerators, washing machines and cars exist, some complain that they are only made available to those people supported by small trade union committees. The danger of favouritism and corruption is quite high. Even so the Cuban revolution does seem to have benefited many people, and to be a genuinely popular one except to those who lost a great deal of wealth or influence. A combination of things, such as allowing dissidents to emigrate, being threatened by the USA, attempting to produce greater equality for all, genuine achievements compared with the rest of Latin America, and the personality and widespread support for Fidel Castro, all sustain the social and economic development.

1 How successful do the Cubans seem to have been in turning economic development into improvements in social well-being? Do you agree with the social priorities chosen? Say why.

2 What are some of the disadvantages of living in a socialist society?

A widespread network of 'polyclinics' like the one below have ensured that health standards are high. (Below left) a meeting of patients and nurses at a maternity hospital

BOLIVIA
1 Environment and history

For large-scale map, see page 123

The cold, arid atmosphere of the altiplano is ideal for drying meat

Physical environment

The Republic of Bolivia is a huge land-locked country in South America, some five times the area of Great Britain, or as large as France, Spain and Portugal combined. Within it are great contrasts of environment. The western border with southern Peru and northern Chile consists of the ice-covered mountain ranges of the Andes known as the Cordillera Occidental. To the east are similar but lower Andean ranges. Between these mountain ranges lies the high plateau known as the altiplano. This bleak area is about 400 kilometres (250 miles) long and is at an altitude of 3600 metres (12 000 feet) above sea level – that is almost three times higher than the top of Ben Nevis, Britain's highest mountain! At this height the air is clear and dry, and the plateau is windswept, treeless and cold. In the north of the altiplano lies Lake Titicaca, the world's highest navigable lake, shared with Peru. Over half the population live on the altiplano, almost 1 million of them in La Paz, the country's largest city.

Another distinctive region consists of the deep valleys and gorges to the east of the mountain ranges enclosing the altiplano. These lush valleys, known as

Ferries on Lake Titicaca, the highest lake in the world

yungas, lie between altitudes of about 1800 and 2750 metres. Further south the valleys are drier, but are densely settled and contain towns such as Cochabamba and Sucre. To get from La Paz to the nearest of the yungas, little more than 40 kilometres away, it is necessary to climb 900 metres to the mountain pass and then drop 1500 metres down the eastern side. Land communications across Bolivia are difficult.

The largest region of the country, occupying about 70 per cent of the total area, consists of tropical lowland plains, varying in altitude from 450 metres on the edge of the foothills to 90 metres in the east and north. The area is drained by headwaters of tributaries that flow northwards into the Amazon and southwards into the River Paraguay. Some of this lowland is covered by tropical jungle while other parts are open forest and savanna grassland used for grazing livestock. Santa Cruz, with a population of just under 400 000, is the main urban area in this newly-developing eastern lowland region.

A settlement in the yungas

The people and their history

In 1986 the population of Bolivia was estimated at about 6.5 million. Roughly 70 per cent are Indians, descendants of the original inhabitants of the high plateau. Many of these speak only Quechua or Aymara, and not Spanish, the official language of the country. About 5 per cent are of European, mainly Spanish, origin. The remainder are people of mixed race, known as mestizos. The great majority of the population are Roman Catholics, reflecting their past history and conquest by the Spanish. Just under half are urban dwellers living in the larger cities or in the newer towns of the oil-rich eastern lowlands.

The altiplano was a centre of very early Indian culture, and became part of the Inca empire around the year 1200. The area fell to the Spanish invaders in 1538. The Spaniards were interested in acquiring wealth, particularly silver from the huge deposits around Potosi, and in converting the people to the Christian faith. Cities such as La Paz and Sucre were founded during the early years of Spanish settlement. The struggle against Spanish rule began early in the nineteenth century and independence was achieved in 1825. At that time Bolivia had access to the Pacific Ocean, but lost the province of Atacama to Chile during the war of 1879–84. Bolivia also lost territory to Brazil in 1903 and to Paraguay in 1935. As with many countries, the frontiers of today are the result of past conflicts with neighbouring states.

Bolivia is a democratic country, and according to its constitution can elect a President and government for a four-year term of office. But the country suffers from a considerable amount of political uncertainty, and at times in recent decades elected governments have been replaced by military rule, based on the power of the armed forces. Elections were held in 1978, 1979 and 1980, but were declared invalid because of fraud, and military rule continued

until 1982. Civilian government was restored in that year and successful early elections held again in 1985. There is strong and organised workers' opposition to the government's economic policies, but no prospect of a political left-wing take-over at the present time.

1 Suggest some of the advantages and disadvantages for economic development of the position, size and varied environments of Bolivia.
2 Name other countries that have been colonised by Europeans. In what ways have the original peoples of Bolivia gained and lost through Spanish conquest and colonisation?

A bus in La Paz. La Paz is the highest capital city in the world

2 # The altiplano

The altiplano is a windswept, almost treeless area. The main vegetation is a coarse grass which provides grazing for llamas, alpacas and sheep, and until recently was the main material used for fuel and thatching the roofs of the dried-mud houses. Rainfall is unreliable, but is usually sufficient for crops to be grown in the northern part. Ancient terraces dating from before the Spanish conquest suggest that the altiplano has been farmed since early times.

When the Spaniards colonised the altiplano in the sixteenth century they divided the area into hacienda, or very large estates. The dispossessed Indians were forced to labour from dawn to dusk for absentee landowners in return for the right to work tiny plots of land. On these they grew their traditional crops of potatoes, wheat and local vegetables, managing to survive under these harsh conditions in extreme poverty. They rarely set foot outside the hacienda, except to transport estate produce to a local town or to La Paz, or to attend a fiesta. Homes consisted of mud adobe huts, thatched with grass, in which lived families and sometimes animals. There was no clean water or sanitation. Hygiene was poor, infant mortality high and life expectancy low. Life for most campesino peasants was hard and miserable.

Land reform

The government that came to power in 1952, supported by the Indians and mineworkers, was devoted to a programme of reform. Amongst other aims it gave high priority to land reform. The large estates were compulsorily split up and redistributed amongst the campesinos and their communities. Before the revolution 95 per cent of the farmland was controlled by about 4 per cent of the wealthiest landowners. The aim was to abolish this system and give the former estate lands to tenant farmers

At the weekly market in the village of Huatajata, 'campesinas' (wives of peasant farmers) buy dyes from Peru

Co-operation over ploughing among peasant farmers on the altiplano

A new home on the edge of Lake Titicaca (above left) and a sign outside a new settlement advertising the fact that all the homes have piped water (above). As a result of developments like this, health standards are rising fast

and labourers who had previously worked them. In some areas the peasants had occupied the estate lands in the years just before the reforms. Land reform was well-intentioned, but land redistribution did not always work well. Some estates were unevenly divided because hacienda labourers got control of their previous plots regardless of their size or quality. There were also many disputes between former landlords and tenants, between peasants and government officials, and between communities and families themselves over who had what rights to what land.

Other problems arose because of the size of the new farms. It is estimated that on the altiplano a family of four needs 3–4 hectares of land to produce enough food and grazing for basic subsistence. In practice some families received less than 1 hectare. In addition to land shortage, many peasant farmers had no experience of efficient farm techniques. Most were without proper tools, and knew little about storing and marketing their produce or obtaining credit. As a result some of the more go-ahead began acting as middlemen, transporting produce to markets, selling essential items to the farmers and acting as agents. Inevitably this led some into debt and exploitation.

On the other hand many families and communities benefited from the reforms. This has been reflected in the building of new homes with more rooms, the provision of electricity and piped water, and the beginnings of better services and markets in local towns. Goods are brought in from La Paz, and farm produce such as coffee and bananas from the yungas further east. Roads have been improved and rural buses have replaced open lorries as the main means of transporting farmers and their families to markets. Some areas, particularly around Lake Titicaca, have become centres for weekend trippers from La Paz and for foreign tourists.

Education is increasingly seen as an important way into jobs in teaching, the police force and the army, while a knowledge of Spanish is essential for successful trade. In particular there has been a change in attitude towards schooling for girls. Women have also become more involved in local businesses such as marketing eggs, onions and other farm produce, and making and selling woven items to tourist shops. The increased independence of women, who in the past were strongly dominated by the men, is one of the major results of the reforms. Women are now better educated, more active in business and more modern in their style of dress, and they are more self-confident and outspoken about their place in the family and the community. While there are signs of development much poverty remains, and a great deal of improvement is needed in such areas as hygiene and health care before the aims of the 1952 reforms are fully realised.

1 a) Name another country where land reform has been carried out. b) What are some of the arguments for and against such compulsory change?
2 Why are improvements in the status, influence and rewards of women an important feature of development? Why are such improvements often resisted and hard to achieve?

183

3 The eastern lowlands

Long before the Spanish conquest and colonisation of the eastern lowlands of Bolivia, Indians successfully lived in the area. Some were hunters and gatherers, but a large number were farmers living in large villages scattered throughout the rainforest and drier savanna woodlands. Their ways of life were disturbed by the Spanish colonists and later on by business interests wanting to exploit the forest resources, particularly rubber and quinine. Some of the lands were also cleared to provide grazing land for cattle. Missionaries also contributed to the change in traditional practices of the lowland Indians. Partly as a result of widespread disease through contact with the European and mestizo settlers, partly through violent conflict, many hundreds of thousands of Indians are estimated to have died during the early colonisation of the eastern lowlands. It is estimated that between 25 000 and 125 000 Amerindians remain in this huge region of Bolivia.

A migrant family from the altiplano arrive at a new village

Economic development

In the mid-1950s, amongst its many problems, the government was faced with those of over-population on the altiplano farmlands, and of large payments for food imports. It also needed to earn foreign money by exporting goods. The eastern lowlands were thought to provide the key. Attention was focused first on the Santa Cruz department and then those of Beni and Pando to the north. The intention was to grow crops such as rice, oil-seed, sugar and cotton in cleared areas. Labour would be provided by the permanent migration of people from the altiplano and the over-populated yungas.

During the 1970s these provinces became new areas of agricultural wealth through rice, sugar, cotton and cattle production. The most important area is north of the city of Santa Cruz. Large farms extend for about 80 kilometres on either side of a highway linking the city with the smaller towns of Montero, Gen! Saavedra and Mineros. Beyond these estates are many smaller farms on which live many of the quarter of a million migrants to the eastern lowlands. Most of the peasant farmers of the altiplano do not want to leave their homes for such a different environment, but many thousands have moved to these less densely settled parts to make new permanent homes on land that will be their own.

Migrant labour

A large amount of the new farmland is devoted to cotton growing, and cotton is now an important export crop. Hand-

A cross-section across Bolivia

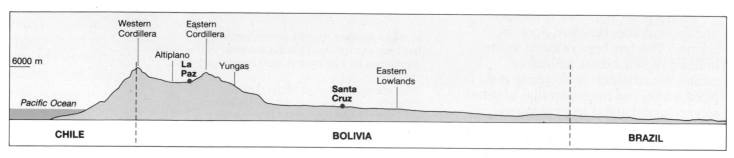

picked cotton commands a higher price than machine-picked, but it also needs cheap labour. Cotton harvesting is a seasonal job, so workers are needed for only a short time each year. The job of finding cotton harvesters is usually done by middlemen, with up to 200 pickers being needed by a large plantation. The middlemen recruit these in the altiplano or yunga villages, and take them by truck to the plantation. Here the workers, who may be complete families, live in shacks or communal huts with very little comfort or medical support. It is hot and humid, quite unlike the conditions of the plateau and yungas. These migrant workers who arrive and leave each season are not well organised, and wages are low. They will earn enough cash to enable them to buy the things they cannot grow or make themselves back home. Most have to repeat the migration year after year in order to survive. Migrant labour is a reflection of the high rural unemployment and underemployment that exists in many parts of the country.

The growth of the cocaine crop

One of the major growth areas in agriculture is the coca plant. The leaf has been chewed as a stimulant by the Indians from time immemorial, but it is now used to make the drug cocaine. This is in huge demand in the USA and elsewhere. In 1983 some 100 000 farmers grew the crop, and produced an

A forest area on the eastern lowlands now cleared for ranching

estimated 48 000 tonnes. Of this 12 400 tonnes were legally grown and sold, 570 tonnes were used for medicinal purposes, and the remaining 35 730 tonnes were illegally grown and smuggled out of the country, often by light aeroplanes. Attempts have been made to stamp out the illegal production, and armed troops have moved into the main producing areas. The USA gives money to try and persuade the farmers to grow alternative crops, but the income from growing the illegal crop makes this difficult. Needless to say the great bulk of the profit is made by the drug dealers and traffickers, not the peasant farmers.

1 In what ways can changes in the eastern lowlands be said to be a) economic and b) social development?
2 If the export of cocaine is producing so much income for the peasant farmers and foreign earnings, why should it be stopped?

Picking coca leaves on the terraces of a yungas valley

Bolivia cocaine racket threatens democracy

Bolivia's cocaine racket has become so vast and economically potent that it threatens the stability of democracy.

President Estenssoro said 'traffickers are so wealthy and powerful that soon they could buy an election.'

Bolivia is considered the world's main producer and exporter of illicit cocaine, produced from the coca leaf, chewed for centuries by peasants and miners of Andean high plains for its mild narcotic effect.

Local and American drug enforcement officials estimate that each year up to $5 billion worth of

cocaine is exported from production centres in the central Cochabamba valley and the far north of the Beni and Pando provinces.

Up to £800 million of that returns to Bolivia, while the rest goes to numbered bank accounts in Europe.

President Estenssoro said Bolivia needed international aid in the fight against the cocaine kings.

The US provides about $3 million a year in anti-drugs aid, and for a voluntary coca plantation reduction programme. It has been a notable failure.

4 Continuing problems

In spite of important reforms in recent decades leading to land reform, voting rights for everyone and nationalisation of much of the mining industry, Bolivia remains one of the poorest countries in the Western hemisphere in terms of production per person. This is partly due to it being so dependent on foreign trade, with a heavy dependence on the export of tin and other minerals. The international price of tin has slumped over recent years, causing a big drop in export earnings and adding to Bolivia's economic problems.

Tin and natural gas

About 80 000 people are employed in the mining industry. This is only 4.5 per cent of the workforce, but the mines do provide about a quarter of the government's income and almost half of the country's export earnings. Some of the larger mines are owned and run by the state. The state mining company Comibal produces over 60 per cent of the country's lead and zinc, 90 per cent of its silver, and almost all of its copper. In the first half of the 1980s productivity declined by almost a quarter due to a combination of exhaustion of higher-grade ores, lack of investment in the industry for exploration and new developments, and many strikes and disputes over low wages and poor working conditions. Equipment is out-dated and safety precautions hardly exist. Wages and conditions are slightly better in some of the medium-sized private mines, but are generally appalling in the many smaller ones. Living and working conditions for the miners and their families are very bad. Life expectancy for the miners is between 30 and 35 years due to industrial diseases, while the mining townships lack basic housing, health and education amenities. There are large smelters near Oruro and Potosi, but much of the ore has to be exported for smelting elsewhere.

There are oil reserves in the southeast of the country in Tarija and Santa Cruz departments, but Bolivia is not a major oil producer. On the other hand natural gas production is important. Argentina is the main foreign customer, and is supplied by a 526 kilometre (377 mile) pipeline from the gas fields to Yacuiba on the Argentinian border.

Bolivia's export and import figures

Value of export 1983 (%)

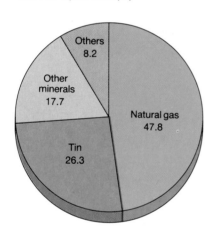

Others 8.2
Other minerals 17.7
Natural gas 47.8
Tin 26.3

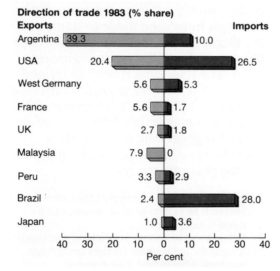

Direction of trade 1983 (% share)

	Exports	Imports
Argentina	39.3	10.0
USA	20.4	26.5
West Germany	5.6	5.3
France	5.6	1.7
UK	2.7	1.8
Malaysia	7.9	0
Peru	3.3	2.9
Brazil	2.4	28.0
Japan	1.0	3.6

Per cent

Bolivian mineworkers live in settlements like this and experience conditions of great hardship

Manufacturing and communications

There is not a very large manufacturing industry in Bolivia, although factories employ about 10 per cent of the work-force. Industrial development is hampered by the very small domestic market – there is little money within the country to buy goods. The main products are foodstuffs, drinks, tobacco and craftwork. On the industrial as opposed to consumer side there is some refining of the oil and various minerals produced within the country. Most capital and consumer manufactured goods have to be imported, which is a drain on the country's resources.

Being without a seaport and depending on international trade, Bolivia spends a lot on transport, storage and customs charges. The country has three international airports, at La Paz, Cochabamba and Santa Cruz. The road system is being improved, and there are highways between La Paz and Cochabamba and Oruro, and from Cochabamba to Santa Cruz. The completion of the latter in 1954 was important in opening up the eastern lowlands and bringing the different regions together. Internal air transport is an important means of linking the highlands with the lowland areas. Apart from the highways a relatively small percentage of roads are properly surfaced for heavy traffic. There are only just over 3500 kilometres (2200 miles) of railway track in this huge country. Poor communications and an underdeveloped transport system certainly hinders development.

Inflation and living standards

The country has had to borrow money from foreign banks and countries, and with big increases in interest rates has found it increasingly difficult to repay them. In 1984 it would have taken over 120 per cent of export earnings to pay off interest and capital repayments for that year! In order to try and get out of debt, and to gain new loans, the government has been forced to cut imports and stop wage increases. It has devalued its money to help, with the result that prices have risen, particularly for imported goods, and wages have fallen

La Paz is growing rapidly and experiencing all the modern urban problems of congestion and pollution

behind inflation. Consumer prices increased by about 20 per cent in 1979. By 1984 this had risen to 1682 per cent. It is no wonder that in 1985 and 1986 workers went on strike against the harsh austerity measures. An indefinite state of emergency was declared in September 1986 by the government because of what it saw as a threat of civil war. In spite of many reforms and its very considerable resources, Bolivia exists in a precarious economic and political situation.

1 Describe some of the advantages and disadvantages to Bolivia of being part of an international economic system.
2 What alternatives might be available to any government attempting to improve the economy of Bolivia?

The rate of population increase is a constant threat to improvements in living standards. And yet children like this deserve to expect more in the way of health and education than their parents

Part Three   Exercises

Forty-five pages of exercises

Each page of exercises in this part of the book links with the corresponding double-page spread in Chapters 1–4 of Part One.

1.1 Similarities and differences

There are great differences in ways of life, living standards and quality of life around the world. Some of these differences contribute to the richness and pleasure of peoples experience, but others reflect injustice and misery, and need to be changed.

These photographs have been chosen to illustrate the theme 'similarities and differences'.

1 For each pair of photographs give a detailed description of what can be seen, and what is similar and different in each.

2 From the illustrations in this book choose two from Asia and two from South America. One photograph from each pair should show a feature, event or behaviour that you might see in your area, the other should show something very different.

3 What types of similarities and differences between places cannot be effectively illustrated in a book?

Boys at play: in Britain (top) and in Mozambique (bottom)

A wedding: in Britain (left) and in India (above)

Even with the best of intentions, pictures and words are bound to be biased and give only a partial picture of the real situation or explanation.

These three pictures were taken in India.

1 Write a paragraph about each picture, describing what it shows. Compare your descriptions with someone elses. Note **a**) the things that you have both mentioned, and **b)** the things that only one of you has mentioned. For each example, say which is the 'best' description, and why you think so.

2 What do these pictures, and others on pages 7, 14–17 and 68–69, tell you about work and housing in India?

3 It is usually impossible to use all the pictures that are needed to give a full account of a place. What impressions have you gained of life in India? What other images of India have you seen? Consider contrasting images, such as travel and holiday advertisements.

1.2 Bias in pictures and words

1.3 Bias in maps and statistics

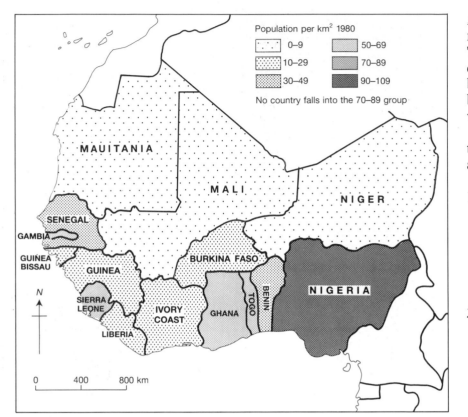

(Above) West Africa:
population density by country

As with pictures and words, so with maps and statistics, bias is inevitable. This may be in either the selection of data, or in the way it is presented, or both. In some cases data is deliberately biased to stress a particular point of view.

These two maps show certain features of the population of West Africa as a whole, and of Nigeria.

1 What information about the population of the region and the countries is given by the map of West Africa? What information about the population of West Africa is obscured or not given? How useful is this type of population map?

2 In what ways is the population map of Nigeria a better one to show features of population in that country? What other information would it be helpful to know? What are the weaknesses of this map?

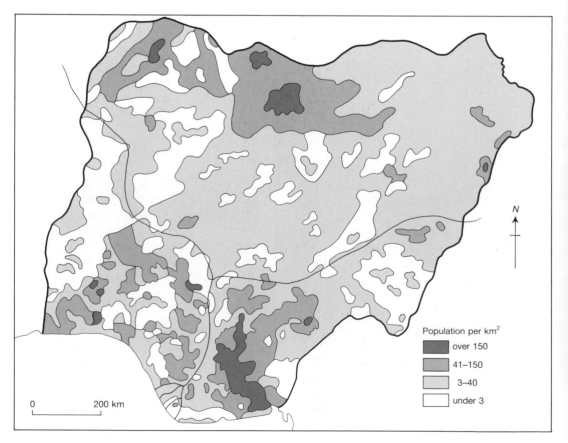

(Right) population distribution
in Nigeria

One of the great differences between people is the work they do and the skill and effort that goes into it. Another is the difference in rewards for that skill, responsibility and effort.

1.4 Work and wealth

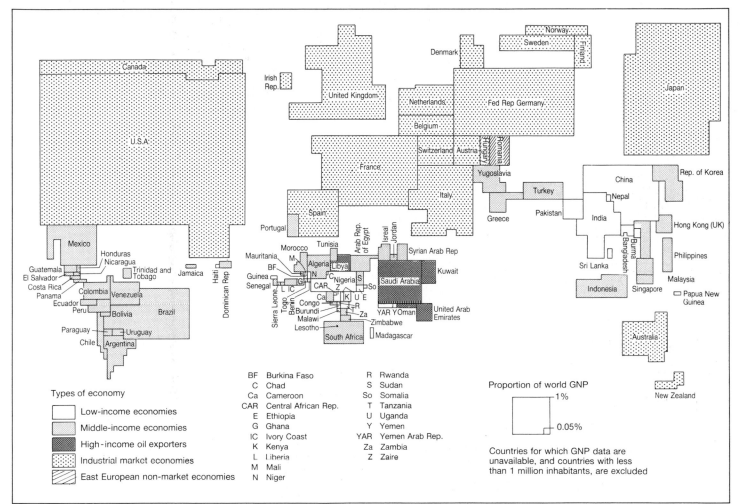

Types of economy

- Low-income economies
- Middle-income economies
- High-income oil exporters
- Industrial market economies
- East European non-market economies

BF	Burkina Faso	R	Rwanda
C	Chad	S	Sudan
Ca	Cameroon	So	Somalia
CAR	Central African Rep.	T	Tanzania
E	Ethiopia	U	Uganda
G	Ghana	Y	Yemen
IC	Ivory Coast	YAR	Yemen Arab Rep.
K	Kenya	Za	Zambia
L	Liberia	Z	Zaire
M	Mali		
N	Niger		

Proportion of world GNP — 1%
— 0.05%

Countries for which GNP data are unavailable, and countries with less than 1 million inhabitants, are excluded

In this world map the shapes of the countries have been ignored, but they are shown more or less in the correct position in relation to others. The size of the country shapes does not represent the area of the country, but the amount of wealth it created during a year in the early 1980s.

1 Using this map and a more conventional map of the world, name
   a) a very large country with a large percentage of the world's wealth,
   b) a very large country with a small percentage of the world's wealth,
   c) a very large country whose wealth is not given,
   d) two places (one is not an independent country) in the Far East that have greater shares of the world's wealth than would be expected from their areas.
   e) a country in the Caribbean that has a East European-type non-market economy and is not shown on the map,
   f) a European country with a middle-income economy.

2 a) What is useful about this type of map for showing the wealth of countries? b) Even if every country were shown, why is the title 'The world according to GNP' a misleading one to use?

1.5 Living standards

The different wealth of people shows in their ability to afford adequate food and water, education, health care and housing – the basic needs of all people.

The extract looks at basic needs in the Third World.

1 For each of the six basic needs described in the extract draw a simple statistical diagram or statistical map to show the information.

2 Which Third World area seems to have the greatest problems?

3 Why does such information about contrasts in living standards have to be interpreted with caution?

Basic needs – the bottom line

For most people in the developing countries the first worry is not the rupee or peso wisely spent but whether they'll have enough to get by. The basic needs for food, clean water, housing, education, health and work are for many people still largely unmet.

Food

Out of a total world population of about 4 500 million, 500 million are actually starving, most of them children under 5 years.
Source: UN FAO

Water and sanitation

Over half the people (1 320m) in the Third World (excluding China) do not have clean water and 1 730m – three-quarters – lack adequate sanitation
Source: Earthscan

Infant mortality

Average infant mortality per 1 000 live births in developing countries is 86, but in Africa it is 120. In the West, the rate is 18 per 1 000 live births.
Source: US Foreign Policy and the Third World 1983

Housing

Nearly half the inhabitants of Third World cities cannot afford the cheapest house which meets basic health and safety needs. More than one million live in illegal shanties in the major cities like Lima, Lagos, Delhi, Cairo and Mexico City.
Source: UN Centre for Human Settlements

Literacy

Average literacy in developing countries is 57% but for Africa it's only 36% whereas in Latin America it's 79%. In low income countries 90% of boys aged 6–11 are in primary school but only 64% of girls.
Source: US Foreign Policy and the Third World 1983 & State of the World's Children Report, 1985

Life expectancy

In the Third World, average is 57 years, going up to 69 for Oceania, 63 for Latin America, 58 for Asia but only 49 for Africa.
Source: US Foreign Policy and the Third World 1983

New Internationalist, May 1985.

Quality of life is affected by material standards of living, but also by such things as stress, anxiety and fear.

Read the extract describing the differences between various groups of people in India and the attempts of one state to create fairer economic and social conditions for the least well-off.

1 Which groups are being given special help in getting jobs and college admissions?

2 What are some of the reasons for giving advantages to such disadvantaged groups? Why are some politicians using these laws for their own ends?

3 Who protested against such help being given, and why? How did the 'backward classes' react when the proposals were changed?

4 Do you think that disadvantaged people should be favoured or helped in getting jobs and social services by 'positive discrimination'? How are disadvantaged groups helped in the United Kingdom and the EEC?

This Harijan family, with their European-breed cow, are part of a scheme to integrate the scheduled castes

---

India

Help for disadvantaged classes may cause irreparable damage

The policy of positive discrimination in favour of economically or socially backward classes is supposed to help social integration and remove inequalities.

Instead, the policies are causing riots and political disturbances, and may eventually damage the social fabric of certain states beyond repair.

The policy calls for places to be reserved for applicants from specified castes or other groups in higher education or in government employment.

It is seen as a useful tool in overcoming prejudices. Lately, however, it has become a way for politically powerful groups to buttress their own privileged positions, encouraged by vote-hungry politicians.

Mr Rama Rao is chief minister in the southern state of Andhra Pradesh. In July 1986, feeling threatened by a move to unite the backward classes against him, he hastily increased reservations in jobs and college admissions.

After his move the number of places available to open competition fell to a ludicrous 29 per cent. The others are divided 16 per cent for scheduled castes, the so-called Untouchables or Harijans; 6 per cent for scheduled tribes, that is the abori-ginal inhabitants of the subcontinent; another 6 per cent for special groups such as the handicapped, and an astonishing 44 per cent for other backward classes, which Mr Rama Rao had increased from 25 per cent.

Not unnaturally, the higher castes were aghast: an important protest began to roll, led by a student agitation. But before much steam could be generated, the Andhra High Court struck down the decision as unconstitutional, citing a Supreme Court decision that reservations should never be more than 50 per cent of the available places.

Mr Ramo Rao agreed with the students not to appeal against the High Court Decision, if they agreed to call off their protests.

But this deal enraged the leaders of the backward classes, who began a series of riots of their own, burning rather more than £2 million worth of buses in the course of it.

1.7 The Third World

The least developed countries (LLDCs) 1984	Population (millions) 1982	Av. Annual pop. growth (%) 1970–82	% urban 1982	Per capita GDP ($) 1982
Benin	3.62	2.5	26	288
Botswana	0.97	3.8	16	824
Burkina Faso	6.36	1.4	9	185
Burundi	4.26	1.8	4	261
Cape Verde	0.31	1.7	20	278
Central African Republic	2.39	2.1	35	270
Chad	4.68	2.1	18	147
Comoros	0.42	3.7	19	237
Djibouti	0.33	6.4	74	636
Equatorial Guinea	0.37	2.0	54	195
Ethiopia	32.78	2.4	14	137
Gambia	0.63	2.9	18	357
Guinea	5.06	2.1	19	404
Guinea-Bissau	0.85	4.1	24	191
Lesotho	1.41	2.4	5	240
Malawi	6.27	2.6	8	213
Mali	7.34	2.1	17	157
Niger	5.61	2.6	13	354
Rwanda	5.51	3.3	5	251
São Tomé and Príncipe	0.09	1.5	—	331
Sierra Leone	3.41	1.6	25	375
Somalia	5.08	5.1	30	285
Sudan	19.79	3.0	21	462
Tanzania	20.23	3.4	13	247
Togo	2.68	2.4	17	307
Uganda	14.12	3.1	7	229
Afghanistan	16.79	2.6	16	232
Bangladesh	92.59	2.6	11	115
Bhutan	1.33	2.0	4	102
Lao PDR	4.10	2.6	13	103
Maldives	0.16	3.0	21	360
Nepal	15.37	2.5	5	153
Yemen AR	7.43	2.9	12	382
Yemen	2.09	3.2	38	410
Haiti	5.20	1.7	26	315
Samoa	0.16	1.0	21	532

Source: UNCTAD *Least Developed Countries 1984 Report*

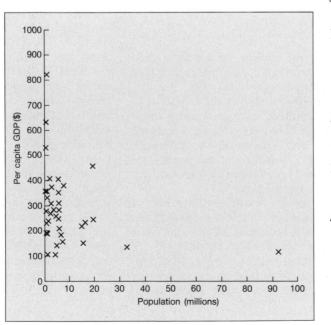

Scatter graph of population against per capita GDP

There are many different ways of grouping the regions and nations of the world, each method having some strengths and some weaknesses. In one such system the relatively poorer and disadvantaged countries are known as the Third World. It is essential to realise that there are great differences of living standards and quality of life within the so-called Third World.

The poorest countries of the world have been described in many ways. The World Bank uses a classification based just on GNP per capita, and in their report of 1984 identified 34 'low income countries'. These countries represented roughly half the world's population.

The United Nations uses the classification of 'least developed countries' (LLDCs), based on a wider range of criteria: low per capita income, low literacy rates (less then 20 per cent of population aged 15 or more), and low contribution of manufacturing to GDP (less than 10 per cent). In 1984 they identified 36 LLDCs. The combined population of these countries was 300 million, or 6.6 per cent of the world total.

Only 22 countries fall into both classifications. And some countries that can be described as belonging to the Third World do not fall into either classification, such as Indonesia, Nigeria and Bolivia.

The distribution of the LLDCs is shown by the map on page 22.

1 Look at the table, and the map on page 22. Which parts of the world include many least developed countries? Which Third World areas include few of these countries?

2 Which LLDCs had over $400 per capita GDP when the report was made?

3 Does the scatter graph reveal any relationship between population and per capita GDP?

4 Draw a scatter graph of either average annual population growth or percentage urban against per capita GDP to see if there is any relationship between them.

2.1 The drylands and food production

Large parts of the world are extremely dry or have unreliable rainfall. It is very difficult for such environments to support large populations if they are dependent on the food they grow themselves. On the other hand some people have learned to live in harmony and balance with such arid environments.

These exercises relate to the climatic conditions of Northern Africa.

1 Study and/or draw graphs of the climatic statistics for the five places named and shown on the map. Give the correct name and number for each of the five places.

2 Give a brief description of the location and features of the climate of the two places that could be included in the 'drylands' of Northern Africa.

The climatic statistics are for the five places shown on the map

The places are: Ilizi
Lagos
Tunis
Agadez
Kano

Place A (Altitude 550 m)

	J	F	M	A	M	J	J	A	S	O	N	D	
Temp (°C)	12.5	15.5	19.9	24.7	29.4	33.5	33.5	32.7	30.7	26.0	19.9	14.4	Average 24.7°C
Rainfall (mm)	2	2	2	2	2	<2	<2	<2	2	<2	2	<2	Total 18 mm

Place B (Altitude 481m)

	J	F	M	A	M	J	J	A	S	O	N	D	
Temp (°C)	21.4	23.9	28.2	30.6	30.2	28.1	25.7	25.1	25.9	26.8	24.8	21.8	Average 26.1°C
Rainfall (mm)	0	0	2	8	71	119	209	311	137	14	3	0	Total 872mm

Place C (Altitude 38m)

	J	F	M	A	M	J	J	A	S	O	N	D	
Temp (°C)	26.7	27.5	27.7	27.4	26.7	25.6	24.4	24.3	25.0	25.6	26.8	26.8	Average 26.2°C
Rainfall (mm)	40	57	100	115	215	336	150	59	214	220	77	41	Total 1625mm

Place D (Altitude 4m)

	J	F	M	A	M	J	J	A	S	O	N	D	
Temp (°C)	11.0	11.7	13.4	15.7	19.1	23.4	25.9	26.6	24.6	20.4	15.9	12.4	Average 18.3°C
Rainfall (mm)	70	47	43	42	23	11	1	11	37	56	57	70	Total 466mm

Place E (Altitude 500m)

	J	F	M	A	M	J	J	A	S	O	N	D	
Temp (°C)	20.1	22.7	27.1	30.7	33.0	33.0	31.3	29.8	30.7	29.2	24.4	20.7	Average 27.7°C
Rainfall (mm)	0	0	0	1	6	8	49	78	20	1	0	0	Total 164mm

Salah oasis, Algeria

North Africa

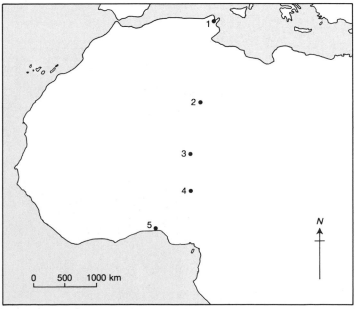

2.2 Drought and famine

These photographs were taken in Ethiopia

Recent decades have seen drought and famine in the arid areas of Northern Africa known as the Sahel. These have been caused by both insufficient rainfall and the degradation of the environment. Environments have been damaged by harmful farming practices and over-stocking, and by the over-cutting of woodlands for fuel. People have, however, often been forced to act in these ways by the need to survive in conditions that are changing beyond their control.

1   Make annotated sketches of these two scenes indicating what they show.

2   Write an account of environmental damage caused by people, based on the two scenes. Try and explain *why* such damage is caused when the damage is so easy to see.

3   What is needed to make it unnecessary for people to cause such damage? What might be done to restore these damaged environments?

The discovery and development of natural resources in some arid areas – particularly oil, mineral ores and underground water – has produced great wealth. In some countries this wealth has been used to improve the living standards of the local people. Other countries, however, have not been able to enjoy such advantages.

In the extract Guntu is the name of a Fulani pastoralist who is involved in one of the League of Red Cross Societies projects in Niger. These involve the siting of wells to provide a permanent supply of water. Guntu also appears in another extract on page 27.

Read the extract and study the map, then answer these questions.

1 What four countries have frontiers with the northern part of Niger?

2 How far is it from the town of Agadez to the frontier with Nigeria?

3 What map evidence suggests Niger is an arid environment? What does the extract say the area was like many thousands of years ago?

4 Why would migrating to nearby towns be 'congregating uselessly'?

5 What seem to be the two main benefits of the sinking of wells in the area? Why are wells likely to be one of the main hopes for secure farming in the future?

Sowing seeds of new life in arid sands of Niger

Guntu's well is one of a group at Tiguerouit near the town of Agadez which sits on the flat compacted sands of the southern Sahara. It is an area of so little rainfall even at the best of times that the dry heat has preserved tiny flint arrowheads from a period 20 000 years ago when the place was lush savanna.

The Tiguerouit project is run by the League of Red Cross Societies which has 21 other schemes in this part of the desert. Diane Hanson, the volunteer who helps administer them, feels they offer a solution to a number of local problems.

'For a start,' she said, 'it has kept people out in the desert which has been their home for centuries instead of congregating uselessly in towns. It is the next obvious step after feeding the starving. It is a small contribution to the revegetation of the desert and forms the base for a return to their old life. As they make some money they can start to build up herds and send their children off with them to find grazing while they remain behind to cultivate.'

But a full re-introduction of the nomadic pastoralism may never be possible. According to Niger's President, General Kountche, the environment is continuously deteriorating here. There may never be the rain to provide any proper grazing yet there are still good underground supplies of water. 'The solution is some sort of combination between agriculture and stockbreeding,' he said.

The old ways of life in the Sahara may have been changed for ever.

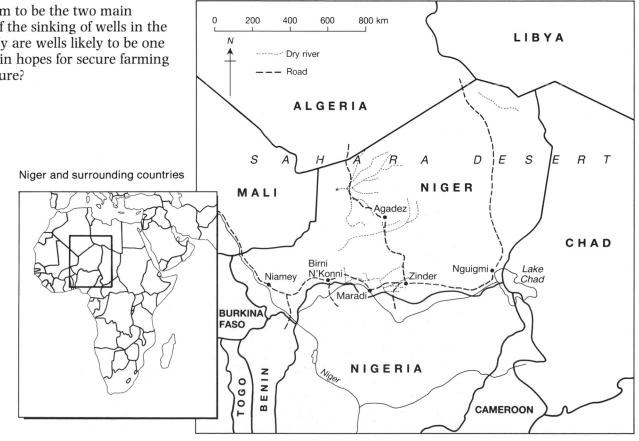

Niger and surrounding countries

2.4 The Arctic

These maps give information about the Arctic region of North America. The route is that taken by the SS *Manhattan*, a nuclear-powered oil tanker specially built for use in the Arctic (photograph on page 32), and the baseline is that drawn by the Canadian government to show what it claims is its territorial waters.

1   Which of the projections is most like the 'real' shapes and areas of Arctic Canada? Why would the Peters projection be very inadequate to show this area? Measure on these maps the distance of the outward route taken by the SS *Manhattan* between Barrow and Thule. What are the difficulties and answers?

2   Why should Canada and the USA want to open up the Artic Ocean to shipping? Which two large US ports would have the shortest linkage via this northerly route? Why should Canada draw a baseline to declare its own territorial waters? If accepted, what affect would this have on US shipping in the area?

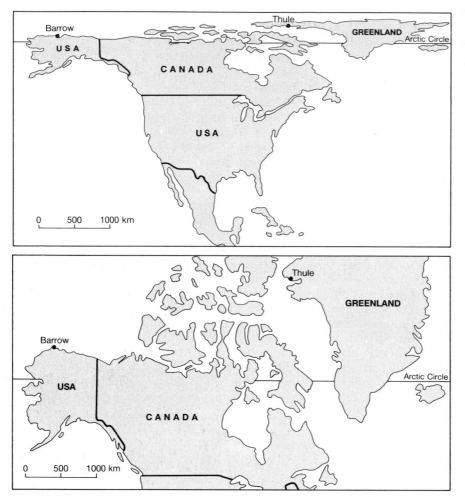

The Peters projection (above, top) and the Mercator projection (above)

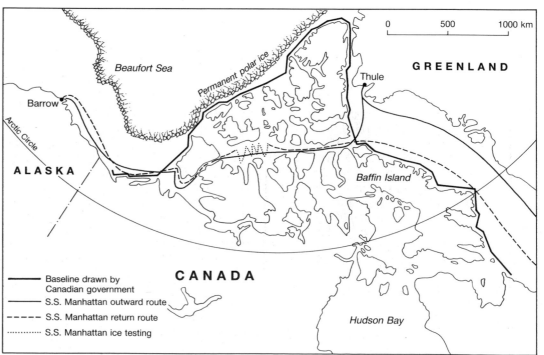

Baseline drawn by Canadian government
—— S.S. Manhattan outward route
- - - S.S. Manhattan return route
········· S.S. Manhattan ice testing

Arctic Canada

The lands fringing the Arctic Ocean provide hostile environments, but people have learned to adapt themselves to survive in such places with great skill and ingenuity.

Most major developments in the Arctic, however, have been undertaken by people from outside these areas. They have acted for their own economic or political reasons and gain. Both the native peoples and the environments have been greatly affected by these resource developments.

The extract looks at the Eskimos or Inuit, their recent experiences, and their hopes for the future.

Read the extract and complete the following exercises.

2.5 Development in the Arctic

1 Describe the environment and wildlife of Arctic Canada.

2 How have Eskimo life-styles changed since the 1950s? With reference to both the extract and pages 30–33, explain the causes of these changes. How would you judge the Eskimo situation as an example of 'development'?

3 What do the Eskimos (or Inuit) hope to gain from Nunavut? In what other ways might they be helped? What do you think the future holds for the Eskimos?

Eskimos hunt for a treeless Nunavut

The 25 000 Canadian Eskimos who live near the North Pole in the eastern Arctic are seeking their own self-governed homeland, *Nunavut*, or Our Land. They would like to remain in Canada, but rule that part of the vast North West Territories that lies north of the treeline.

The temperature is −33°C, and it is the time of the year when there are about three hours of greyish light. We travel across the ice on a sled towed by a snowmobile.

There are no trees, no landmarks, no roads. Yet Alain and Simon drive as though following a road map to the caribou.

After two hours we stop. It seems impossible that anything could exist here. Simon points, and the grey coats of the caribou can just be picked out. Alain unpacks his rifle.

Six are soon killed and the ice is turned into a slaughter-house as they are skinned and cut up. Only hooves, antlers and innards are left for the Arctic foxes. Everything else is piled, freshly frozen, on to the sledges.

The Eskimos possess great survival skills. But they are no longer independent hunters. They have been almost overwhelmed by Canadian kindness, and depend on others for homes, welfare and the few jobs available. *Nunavut* appears to be their best hope of regaining some of the independence they have lost in a remarkably short time.

There was famine during the 1950s. Canada could not let any Canadians starve, and an immense amount of money and effort was spent trying to help the Eskimos. Settlements were established at places like Igloolik, and even if they were far from the best hunting grounds, it all seemed too good to be true to the Eskimos.

Given the choice between their igloos and homes where light and heat came on at the flick of a switch, they did not hesitate. 'It seemed like a never-ending happiness' says the mayor of Igloolik, Lucasi Ivulu. 'No one warned us of the problems we would have in the future.'

Starvation was ended, but no activity replaced the hunt for food. 'In the camps you had to do things constantly, but

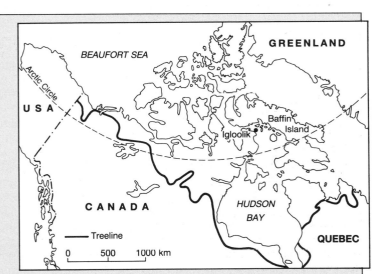

here you sit all day long, waiting,' says Simon's father, André.

The culture disintegrated. By the 1960s the days of independent living were over for the Eskimos.

Now they are beginning to fight back. Sometimes in slightly laughable forms. The night-school in Igloolik teaches Eskimos to build igloos and skin seals. But the quest for *Nunavut* too is part of this fight.

The hope is that *Nunavut* would create more jobs. Unemployment in Igloolik is over 50 per cent and has been aggravated by the Eskimo tradition of marrying young (some brides are only 12) because, in the past, life expectancy was so short. Today Eskimos live longer and more children survive, so the labour market has been flooded with young job hunters.

A homeland would also give them their own language, *Inuktitut*.

What else would their homeland give the Eskimos? 'Our pride,' says the Eskimo MP Thomas Sulluk. 'We would be able to turn round and begin to undo what we have lost in the last 20 years.'

2.6 Wealth from Antarctica

Antarctica is considerably larger than Europe, and covers about one-tenth of the world's land surface. It has never been settled by permanent communities, and only a handful of scientists live there. Recent exploration and scientific investigations have shown that there are enormous resources in Antarctica and the surrounding seas, although 'development' by national governments has been prevented by treaty until 1991. One of the big questions is who should benefit from this – a few more powerful nations, or the whole of mankind.

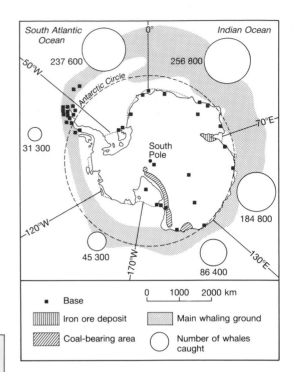

Antarctica

Sitting on top of the wealth

Since 1959 the Antarctic has enjoyed remarkable tranquillity. Russians and Americans, Britons and Argentines, Indians and South Africans, have all been able to shed their prejudices once south of Latitude 60 degrees and work together in remarkable scientific harmony. There has been no military activity, no nuclear dumping or testing, no commercial exploitation, no significant damage to the environment, no reactivation of the various conflicting territorial claims which the Treaty put into abeyance, and an almost total concentration on pure research and mutual support.

But some think the Treaty will be lucky to survive to 1991, when it can come up for renegotiation.

Two large problems threaten its stability – the smaller, less powerful United Nations members' dislike for an important international arrangement which appears to leave them out in the cold; and the question of mineral rights in and around the Antarctic land mass, which is becoming an urgent issue but which the Treaty leaves untouched.

Who will slice the ice?

Natural historian and conservationist David Attenborough says: 'You must bear in mind that the treaty has not really been tested yet. One of the sad things about humans is that they are always very happy to say that everything in the environment will be saved and protected. But in reality it often doesn't turn out quite like that.

'To be specific, if it were to be demonstrated that there were huge deposits of valuable uranium down there under the ice-cap, then for certain there would be someone to stand up and explain why, for the general good, it should be mined. Once this happens, then the floodgates open. The virginity is lost.'

The likelihood is that the long freeze on mineral exploitation will be thawed out by the member nations. One of the most important matters still to be thrashed out is whether resources found in the continent are to be shared equally between the countries, or whether they are to be the sole prerogative of the countries laying claim to the slice in which those resources are found.

The extracts look at the Antarctic Treaty and the issues surrounding it as 1991 approaches.

1. The map provides information about whaling in the early 1980s, but since then it has been greatly reduced by bans and limits imposed by the International Whaling Commission. In 1986 the Commission introduced a world-wide ban on commercial whaling. Why has this been necessary? Why have whales and whaling been the focal point or symbol of environmental and conservation campaigns?

2. Describe other environmental resources available in Antarctica, and some of the technical problems in developing them.

3. List the main achievements of the Antarctic Treaty.

4. Assume that Antarctica becomes the first World Park, to be preserved forever as a vast wilderness. What would need to happen for this to come about, and how likely is this? What would be some of the problems of managing such a World Park, and how might it be used? What are the alternative futures for Antarctica?

2.7 Mountain environments

Large areas of the earth's surface consist of high mountain ranges and enclosed high altitude valleys. Despite the great difficulties and hazards, mountain environments in many parts of the world have been settled for centuries.

1 Give the names of the four major mountain ranges labelled A-D on the map. Describe the locations of the mountain regions numbered 1–8, using the names of continents and countries as appropriate.

2 Using the diagram as evidence, rank the continents according to the altitudes of their four or five highest peaks.

3 What are the problems of development in mountain environments? What are some of the potential resources of such environments for development? Where, and why, have such resources been most exploited up till now?

The world's mountain environments

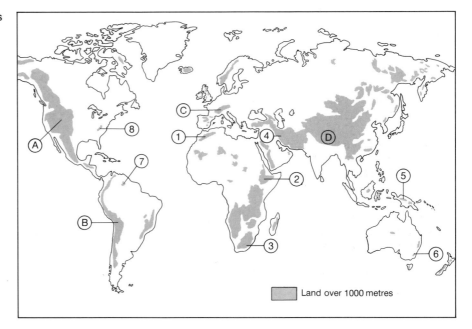

Land over 1000 metres

This diagram shows the highest mountains in each region or continent

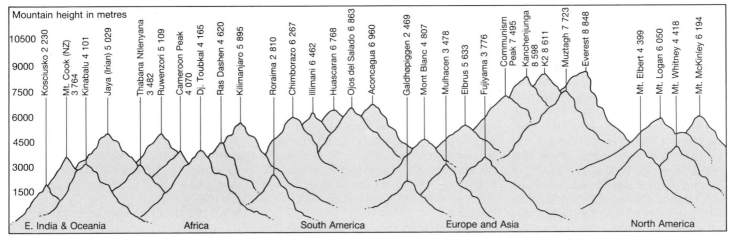

Mountain height in metres

Kosciusko 2 230	
Mt. Cook (NZ) 3 764	
Kinabalu 4 101	
Jaya (Irian) 5 029	
Thabana Ntlenyana 3 482	
Ruwenzori 5 109	
Cameroon Peak 4 070	
Dj. Toubkal 4 165	
Ras Dashen 4 620	
Kilimanjaro 5 895	
Roraima 2 810	
Chimborazo 6 267	
Illimani 6 462	
Huascaran 6 768	
Ojos del Salado 6 863	
Aconcagua 6 960	
Galdhøpiggen 2 469	
Mont Blanc 4 807	
Mulhacen 3 478	
Elbrus 5 633	
Fujiyama 3 776	
Communism Peak 7 495	
Kanchenjunga 8 598	
K2 8 611	
Muztagh 7 723	
Everest 8 848	
Mt. Elbert 4 399	
Mt. Logan 6 050	
Mt. Whitney 4 418	
Mt. McKinley 6 194	

E. India & Oceania · Africa · South America · Europe and Asia · North America

2.8 Change in the Himalayas

Many changes have taken place amongst the communities living in the Himalayan foothills and mountain slopes over the past few decades. These changes have also had a big impact on the environment itself.

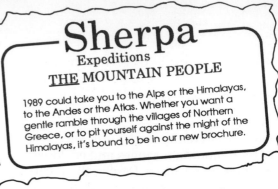

Ever since Mount Everest was first conquered in 1953 the Himalayas have become an increasingly popular area for expeditions and tourism.

1 Copy or trace the map shown in the middle of a sheet of paper. Annotate the route with comments taken from the extract, describing what a) the environment and b) the life of local people is like at various points. Use arrows to indicate the points along the route referred to.

2 What does the extract tell you about tourism in the Himalayas?

3 Look at the advertisement. From what, and why, has the company taken its name? What are the attractions of this type of holiday?

4 What are some of the advantages and disadvantages of tourism for the local peoples?

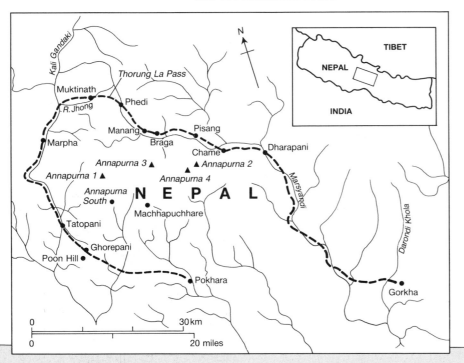

Trekking in the Himalayas

In Kathmandu, we discussed the details of our customised trek – 22 days to circumambulate the Annapurna range, covering some 200 miles – and picked up our Sherpa, Lhakpa, our cook, Jeddha, and six porters to carry the camping and cooking equipment. Starting at Gorkha, on the far eastern side of the Annapurna range, it was clear from the reaction of the people who populated these heavily cultivated foothills of rice paddies and maize and millet terraces, that few Westerners passed this way. Wading almost waist-high through the wide and fast-flowing Darondi Khola and testing calf muscles over a succession of lush hills on paths shaded by huge banana and papaya trees, we met only Nepalese for nearly three days. Our first fellow trekkers were encountered at the suspension footbridge over the Marsyandi Khola.

We followed the Marsyandi northwards, bathing in it and resting by it in the evening. We ascended gradually through pine forests and ticked off the 1 000ft contours on our map. We feared that we might be among those affected by low oxygen in the atmosphere above 8 000ft.

The 17 800ft Thorung La Pass was the choking, freezing, mind-numbing, oath-uttering slog we expected: a scree-covered col surrounded by countless summits under snow. We constantly wanted to sit down for a breather, and were constantly scolded by our Sherpa who knew the fatal consequences of resting in such conditions. Dropping 6 000ft to the religious settlement of Muktinath, the feeling of relief was intense.

We continued down the Kali Gandaki, the world's deepest gorge, and became increasingly depressed about the meagre living conditions of the Nepalese.

Farther on, Tatopani's natural hot springs provided luxurious baths by the river as we watched white monkeys swinging through the oak trees. Then we climbed through the vast rhododendron jungle up to Ghorepani, staging post of Poon Hill and spectacular views of Annapurna South Face's towering glacial pyramid.

Of the 100 or so trekkers at Poon Hill, about four-fifths had arrived via the short, crowded, path from Pokhara. The rest, like us, had been on longer expeditions, arriving along relatively deserted and litter-free trails and without encountering crocodiles of trekkers.

We ended our journey at Pokhara.

The interrelationship between climate, plants and living creatures in lowland areas near the equator has produced the remarkable rainforest environment. In its natural state it provides a very rich variety of flora and fauna.

2.9 Hot, wet environments: the rainforests

1 Name two countries in each of South America, Africa and South-East Asia that include large areas of rainforest. Which two huge rivers flow through the rainforests of South America and central Africa?

2 Put into words the main differences between the climates of the rainforest areas and **a)** the open woodland/savanna areas of northern Africa and **b)** the climate of your home area.

3 Many products of the tropical rainforest are used by people in Britain in their daily lives. Make a list of these to include some foodstuffs, some medicines and some consumer goods. What does this suggest about the variety and value of the rainforest habitat?

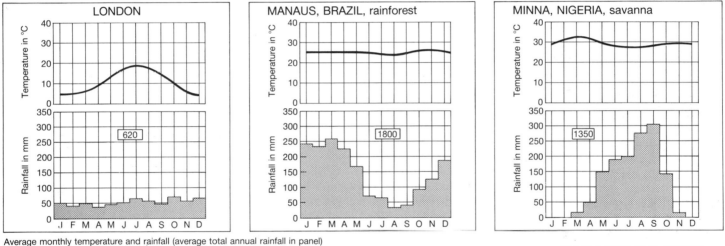

Average monthly temperature and rainfall (average total annual rainfall in panel)

(Above) three very different climate graphs

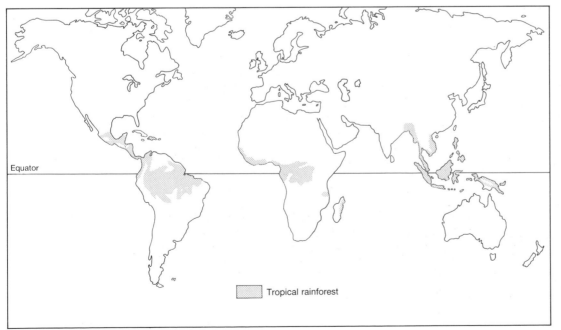

Tropical rainforest

The world's rainforest areas

2.10 Clearance of the rainforests

Recent decades have seen an alarming destruction of rainforests in all parts of the world. There are various reasons for this clearance, and the consequences are causing world-wide concern.

Many conservation groups are trying to make more people aware of what is happening. This advertisement was produced by 'Friends of the Earth'.

1   Write out each sentence in the first paragraph. For each one say whether the statement is unquestionably true, possibly true but needing to be checked, or a matter of opnion.

2   What are suggested as threats to the rainforests? Describe other ways and reasons for the destruction of some rainforest areas.

3   Comment on the 'positive policies' proposed to protect rainforests, saying whether you feel they would be effective. If some possible policies are missing, suggest what they might be.

4   What do you think are the strengths and weaknesses of this advertisement in encouraging and/or ensuring the proper use of the world's rainforests?

How are the mighty falling . . .

Tropical rainforests are the richest resource on Earth. They are home to many indigenous peoples and up to half the world's wild creatures. As a source of medicines, genetic reserves, lasting economic wealth and climatic stability, they are of incalculable value to us all.

Yet they are being recklessly destroyed.

Up to 50% have already disappeared. And an area the size of England, Scotland and Wales is lost every year.

Friends of the Earth is now mounting its biggest ever campaign, internationally, to promote the responsible and sustainable use of the forests, and provide a secure environment for their species and peoples.

Tropical rainforests may seem far away, but the problem is one which concerns us all.

Challenging the mighty

Is this problem insoluble in the face of threats from multi-national corporations and unheeding government departments?

We don't think so.

In the interests of us all, Friends of the Earth is challenging those whose actions are destroying our natural wealth and diminishing our real quality of life.

And, today, our voice is heard.

Positive policies

We are heard, because Friends of the Earth does more than criticise. We propose constructive alternatives such as:
● agricultural grant-aid for conservation minded farming
● a Code of Conduct for tropical rainforest users, and the establishment of an international Tropical Rainforest Protection Fund
● a clean coal strategy to reduce acid rain
● an energy strategy based on conservation, renewables and combined-heat-and-power

No-one is so mighty

The Victorians believed we could 'beat nature'. Today we know better. None of us is so mighty that we will not ultimately suffer if we destroy the world's rainforests.

Friends of the Earth will be campaigning hard to preserve the rainforests and much else that concerns the real quality of our lives.

Will you help us?

Friends of the Earth

2.11 Environmental disasters

Some parts of the world are very susceptible to irregular, short but powerful and destructive environmental disasters. These include earthquakes, volcanic eruptions and powerful storms. They tend to have the greatest impact on poorer communities.

Intense storms sometimes occur in tropical areas. They consist of very low pressure systems with an 'eye' or centre around which whirl winds of tremendous force. There is usually thunder and torrential rain associated with the winds. They originate over the sea and move along 'tracks' before dying out. They can cause great damage to shipping at sea and to coastal areas that they cross. They are known by different names in different parts of the world, as shown on the map.

1   What is rather confusing about the headline to the article dated 4th September 1984, and why might that wording have been chosen? What features of the storm are described in the text and on the map? Compare the strength and effect of Ike in 1984 and Joan in 1970. In which months did they occur?

2   Name the countries and/or describe the areas that may be hit by tropical storms in a) the Third World and b) the North or Second and First Worlds. Why are people in Third World countries generally less able to cope with these storms and the damage they cause?

Storm in Philippines

Typhoon Ike, the worst in 14 years, hit the central Philippines yesterday, with 137 mph winds and torrential rain leaving at least 50 people dead, a dozen missing and thousands homeless.

Government television said 11 vessels, including passenger ferries, sank in stormy seas off the central island of Cebu. Ten fishermen are missing and six other vessels were badly damaged or beached.

The typhoon was the strongest since October 1970 when Typhoon Joan battered the country with 170 mph winds and left 575 dead.

The death toll is expected to rise sharply after communications are restored.

From *The Times*, 3rd September 1984.

Over 300 dead in wake of Hurricane Ike

At least 331 people were killed, 200 of them at a devastated lakeside town, during Sundays' destructive rampage of Typhoon Ike through the Southern Philippines, officials said yesterday.

Provincial authorities in Surigao del Norte, on the north-eastern tip of Mindanao Island, said most of the 200 victims in Mainit town drowned when 137 mph winds churned up lake waters which smashed down houses dotted along the shoreline.

Another 82 people died in the provincial capital Surigao 30 miles away, where journalists reported that the

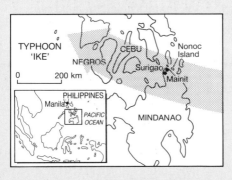

majority of buildings were badly damaged. They said over 300 people were injured.

From *The Times*, 4th September 1984.

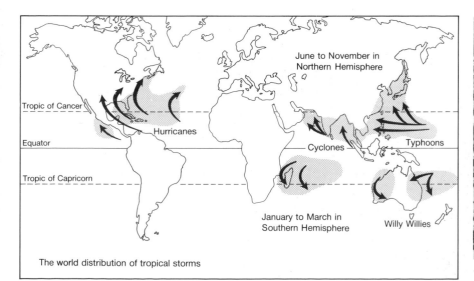

The world distribution of tropical storms

Typhoon damage in the Philippines

207

2.12 Disease

Some people are forced to live in environments that encourage diseases. Very often their ability to resist disease is weakened by malnutrition and lack of medical care and health education.

Combating Malaria

Malaria, transmitted by the Anopheles mosquito, is one of the oldest and most debilitating diseases known to mankind. In many developing countries, especially in tropical Africa, it is one of the five main causes of infant and child mortality. Pregnant women are also especially vulnerable, particularly if already malnourished.

During the late 1950s hopes ran high that malaria would soon be completely eradicated. But the disease has since made a spectacular come-back, largely because of the expense of maintaining control programmes indefinitely, growing resistance to insecticides among the mosquitoes which carry the disease,

and increasing drug resistance in the most important of the malaria parasites.

Long-term control of malaria depends on the selective use of insecticides, eliminating the mosquito's breeding grounds, and using mosquito nets, screens and repellents to protect humans from contact with mosquitoes.

These measures are either expensive, technically difficult, or both. In the meantime, though, mothers and children can be protected against malaria, at low cost, if they are given preventive drug treatment as part of community-based primary health care activities.

Malaria is one of the most destructive diseases that can affect humans. It has led to large numbers of deaths and a vast amount of illness over the centuries. It has been eradicated or controlled in many areas, but still attacks many millions of people and in some parts of the world is on the increase again.

1 Describe in words those parts of the world where malaria may be caught. Which Third World countries are not included? Can you suggest why this is so?

2 Describe and explain the different incidence of malaria on the eight islands named on the map.

3 What problems have prevented the complete eradication of malaria?

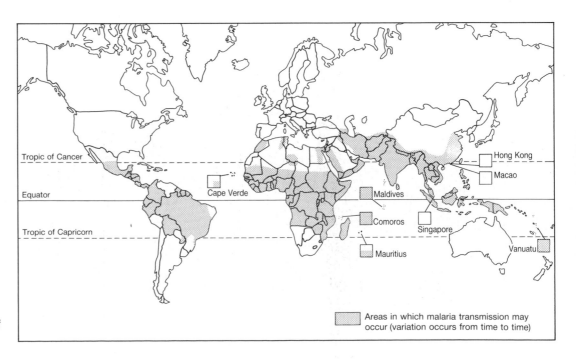

The world distribution of malaria

Areas in which malaria transmission may occur (variation occurs from time to time)

3.1 Population density and living standards

The larger a population is, the greater its need for resources. But there is not a simple relationship between population density and living standards. Some people living in high density areas have a very high standard of living, while in others they are extremely poor.

1 Describe the density of population in the areas shown in each of the photographs. For each, suggest what the standard of living is from the clues provided.

2 From your local area describe **a)** an area of low density and high living standards and **b)** an area of high density and high living standards.

3 When can an area of land be said to be 'over-populated'?

3.2 Population change

The total population of the world is increasing dramatically, although the amount and rate of change varies greatly from place to place. In some places rapid population growth hinders economic and social development. It is often argued that small population growth depends on, or is a result of, economic and social improvements.

Table 1: Population increase

	Population (millions)				
	1950	1960	1970	1984	1980
World	2504	3014	3683	4453	4763
Africa	222	278	357	476	537
North America	166	199	227	252	395
South America	165	217	284	362	406
Asia	1336	1666	2095	2591	2777
Europe (& USSR)	572	639	701	749	765
Oceania	13	16	19	23	24

Every year the world's population increases by more than 70 million people.

Table 2: Rates of population increase

	Average annual population change (per cent)			
	1950–55	1960–65	1970–75	1980–85
World	1.84	1.98	2.03	1.67
Africa	2.11	2.44	2.74	3.01
North America	1.80	1.49	1.05	0.89
South America	2.72	2.80	2.51	2.30
Asia	2.00	2.46	2.35	1.73
Europe (& USSR)	0.82	0.93	0.65	0.35
Oceania	2.25	2.08	1.85	1.50

An average annual growth rate of 3% means that the population will double in 24 years.
An average annual growth rate of 2% means that the population will double in 36 years.

1 Draw a line graph to show the information contained in Table 1. Label each line on your graph. What fact does your graph for Table 1 clearly illustrate?

2 Draw another line graph for Table 2.
   a) What general trend in growth rates is revealed?
   b) Which area remains the exception to this general trend?
   c) Compare the situation in the North and South. Where is most of the world's population growth taking place?
   d) Why might this graph be misleading if viewed just on its own?

3 Look at both graphs. Do you think the population of Africa will ever exceed that of Europe?

4 Look at the map. Name four countries in each category.

5 What sort of economic and social improvements might prompt people to have smaller families?

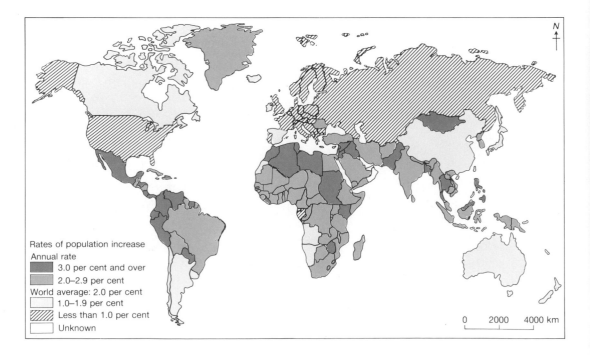

Rates of population increase
Annual rate
▓ 3.0 per cent and over
▒ 2.0–2.9 per cent
World average: 2.0 per cent
□ 1.0–1.9 per cent
▨ Less than 1.0 per cent
□ Unknown

0   2000   4000 km

Rates of population increase, by country

210

Widespread ill-health and lack of adequate health care is a major contributor to lack of economic development in many Third World countries.

The extracts look at immunisation against the Big Six vaccine-preventable diseases. They are taken from the UNICEF report *The State of the World's Children*. The letter considers other factors in Third World child deaths.

1 Identify the Big Six vaccine-preventable diseases. How many children are they killing each year? What is another major cause of child deaths?

2 Read the letter about Third World deaths. What does the writer claim to be 'the main threats to life, by far, …'? Is this view supported by the information in the pie graph?

3 What does the graph state about the relationship between death from diarrhoea, distance from a clinic, and attendance at a clinic, in a part of Bangladesh? How does the information shown by the graph help explain why so many children in the Third World die as described in the letter?

4 What does the letter claim are the main methods of controlling infant mortality?

5 Write a reply to the letter, stating the case in favour of immunisation more strongly.

3.3 Health and health care

Third World deaths

Sir, The claim that the majority of infant deaths in the Third World can be prevented by immunization is a dangerous over-simplification of a complex problem.

It is surely obvious by now that the main threats to life, by far, are malnutrition, gastro-enteritis and dehydration, each aggravating the other and none specifically preventable by immunization.

Third World infant mortality rates were once common in the North. The Big Six were also more common, but they never accounted for the majority of infant deaths and, as causes of death, they fell to very low levels before vaccines were available.

The lesson of history and epidemiology, therefore, is that the control of infant mortality depends on the care of children generally by improvements in personal hygiene, water availability and safety, breast-feeding where practicable, education of parents and older children, with provision of medical or nursing care when and precisely where the need arises.

Immunization is useful, but it is a mistake to believe that it will deal with the greater part of the problem in the Third World.

From a letter by Professor Stewart, Department of Community Medicine, University of Glasgow. *The Times* 4th April 1984

Immunization

Many developing countries face serious supply problems with immunization services. Technological developments are helping to overcome some of these but management capacities need further strengthening.

Immunization is as much a question of demand as supply. Recent evaluations have shown that coverage rates could be doubled and in many cases trebled if parents took advantage of existing immunization services and if those bringing their children for the first vaccination were also to return for the second and third.

● Each year 3.5 million children in the developing world die and 3.5 million more are mentally or physically disabled through six vaccine-preventable diseases: diphtheria, pertussis (whooping cough), tetanus, measles, polio and tuberculosis.

● Vaccines are now saving the lives of approximately one million children a year in the developing world. If the 1990 target for universal immunization is met, then the lives of an additional 3 million children will be saved each year, effectively bringing to an end the reign of the Big Six infectious diseases.

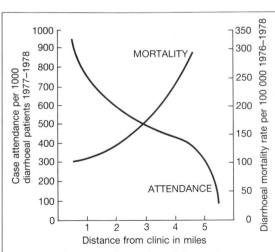

Relationship between distance from clinic, attendance rates, and deaths from diarrhoeal disease (Teknaf Diarrhoea Clinic, Bangladesh, 1977–1978)

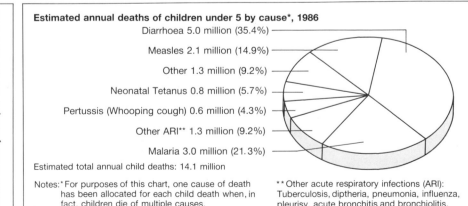

Estimated annual deaths of children under 5 by cause*, 1986

Diarrhoea 5.0 million (35.4%)

Measles 2.1 million (14.9%)

Other 1.3 million (9.2%)

Neonatal Tetanus 0.8 million (5.7%)

Pertussis (Whooping cough) 0.6 million (4.3%)

Other ARI** 1.3 million (9.2%)

Malaria 3.0 million (21.3%)

Estimated total annual child deaths: 14.1 million

Notes: * For purposes of this chart, one cause of death has been allocated for each child death when, in fact, children die of multiple causes.

** Other acute respiratory infections (ARI): Tuberculosis, diptheria, pneumonia, influenza, pleurisy, acute bronchitis and bronchiolitis, otitis media and other respiratory tract diseases.

3.4 Education and schooling

Education is important for both individuals and the future of societies, but its provision is unevenly distributed around the world. There are usually too few resources to provide adequate education in Third World countries.

Adult literacy rate is the percentage of persons aged 15 and over who can read and write. Secondary school enrolment ratio is enrolment at the secondary level as a percentage of the age group corresponding to that level.

1 The adult literacy rate, secondary school enrolment ratio and GNP per capita are given for the countries shown on the map. For countries A–E in the table, say which set of data refers to which country, explaining the reasons for your decision.

2 Is there much difference between male and female adult literacy within each country? Explain why these differences are more marked in some countries than others. Is there any relationship with GNP per capita?

3 Is there much relationship between secondary school enrolment and GNP per capita? Account for your findings. What do you notice about country B? Give an explanation.

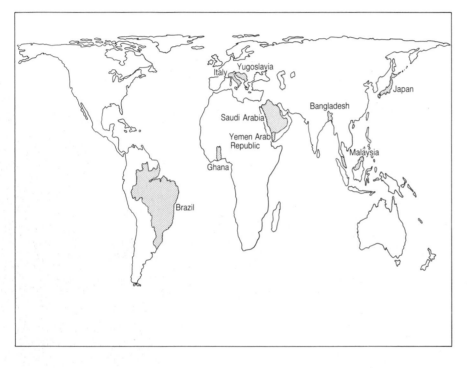

This table provides education statistics for nine selected countries. The nine countries are shown on the map (above)

Country	Adult literacy rate (per cent) 1985		Secondary school enrolment ratio (per cent) 1982 – 1984		GNP per capita 1985 US$
	Male	Female	Male	Female	
A	27	3	16	2	550
B	35	12	42	28	8850
C	79	76	31	36	1640
D	99	99	93	95	11300
E	97	86	84	80	2070
Bangladesh	43	22	26	11	150
Italy	98	96	75	74	6520
Malaysia	81	66	50	49	2000
Ghana	64	43	48	28	380

In many developing countries the educational needs are not the same as those that are provided for by Western-type schools. Third World countries need to provide education appropriate to their particular circumstances.

These photos convey certain messages about life in an African village for young children.

1 List the skills that it is claimed can be learned by children taking part in normal daily life.

2 If schools are not available, what education might such children lack that could be valuable and worthwhile to them and their community? Some suggestions may be got from the diagram on page 58.

Learning everyday...

The photographs show certain learning methods that Western-style education struggles to recreate. Others include:

Home-made entertainment. There is no need for mass-produced toys when you make your own toys and games.

Weaving baskets and decorating pots. No need for 'childrens art' around the classroom wall.

Doing the washing up. No need for nursery-school water trays to learn the properties of water.

Young babies are carried for long periods on their mother's back. Therefore there is no need for rattles to counter the isolation of the pram or cot. It is possible that this accounts for the early development of some African children.

Looking after animals. No need for a pet corner or nature table in the classroom.

Looking after younger children. No need for dolls.

Collecting firewood. Learning about fire and heat, as well as developing manual skills.

Selling vegetables at the market. No need for 'pretend groceries' and 'pretend money' to develop numeracy and basic financial skills.

3.6 Skills for development

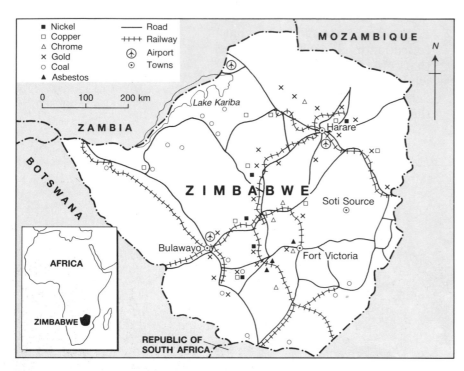

Zimbabwe's transport network and mineral resources. Soti Source, mentioned in the extract below, is also shown

Since the beginnings of time, human beings all over the world have shown remarkable skills in adapting to their environments and developing their resources. All the evidence is that given proper training and practice people in Third World countries could develop skills needed for economic and social activity in the modern world.

The extract looks at the recent life and efforts of a farmer in Zimbabwe.

1 Describe the skills shown by Mr Tazira in developing his farm in Zimbabwe.

2 From the evidence on the map, list some of the skills needed by Zimbabweans to run their country and its economy.

3 Why do most countries in the Third World want to stop being dependent on expatriate or voluntary aid workers as soon as possible?

Independence gives black farmer a new life

Peace and plenty on resettled land

Mr Rashamir Tazira, aged 55, counts himself fortunate. Since the fall of white-ruled Rhodesia in 1980, he has acquired a four-bedroomed brick house which glows bright green with a fresh coat of paint. He has 12 acres of land from which he expects to reap nearly 18 tonnes of maize, and he owns 18 sleek head of long-horned cattle.

His seven children are all at school, four miles away is a clinic, a borehole nearby gushes clear water and his garden blooms with scarlet cannas and canary creeper. What pleases him most, however, is that there are no more weapons around.

In November 1980, a Government lorry dumped him, his family, and 12 other families next to the ruins of a white settler's house, formerly owned by an elderly Afrikaner couple who had been driven off by guerrillas during the Zimbabwean liberation war.

Mr Tazira used to be a shoemaker in the pioneer town of Fort Victoria (now Masvingo). Without money, he returned to his traditional home in Gutu communal land to scrape a subsistence living.

When the Government began its plan to resettle at least some of the hundreds of thousands of people crammed into overcrowded, infertile and hopelessly overworked communal lands, Mr Tazira and his family were the first to be moved. Soti Source, a stretch of rolling hills and valleys adjacent to Gutu, was the first to be resettled.

Mr Tazira arrived too late to make much use of the heavy rainy season of 1980 to 1981, and the next three years were abysmal. Drought followed drought. In mid-1982, he harvested 10 bags of maize, and the next year, he managed to reach self-sufficiency, with 15 bags. But last year the

land yielded 95 bags.

Before independence the biggest crop harvested by peasant farmers was 65 000 tonnes. Last year, the small scale black farmers rescued the country from large scale food imports by growing 380 000 tonnes of maize, nearly 40 per cent of the country's total production.

An exercise book showed Mr Tazira's attempts to budget according to sheets of computer printouts from the Agricultural Finance Corporation.

Repayments from a previous loan were rescheduled, and this season he borrowed Zim$365 (£205) which he used almost exclusively for fertilizers that will enable him to produce 200 bags of maize, more than double last year's crop.

Next year will see the outstanding loan repayments out of the way, more fertilizer, the beginnings of the use of insecticides and a down payment on a cart.

The final step on the road to success for Mr Tazira will be his own farm – as his land is on long-term lease – about 1 000 acres, and a small store.

3.7 Unemployment in the Third World

Unemployment and underemployment is widespread in Third World countries. Some types of development could result in more jobs, but many new technologies require fewer, not more, workers.

The table provides comparative figures for North and South. The photos show unemployed people in the North and South.

1 Draw a statistical diagram to illustrate the figures in the table.

2 What are the likely future trends for unemployment in view of the figures given for children entering the workforce between 1985–2005?

3 What are the differences between unemployment in the North and unemployment in the South? How might a) individuals and b) governments in the North and South respond to the problem of unemployment?

THE NORTH		THE SOUTH
450 million	Employed	700 million
100 million	Unemployed or underemployed	550 million
150 million	'Others' – family labourers etc.	700 million
400 million	Children entering workforce 1985 – 2005	1950 million

Men without work: in London (right) and in a shanty town near São Paulo, Brazil (below)

3.8 Farming

Developments in farming in Third World countries need not only to provide more output, but also work for many people. It is also important to avoid causing damage to the environment.

The map gives information about the position of agriculture in national economies.

1 Name two countries in Africa and two countries in Asia where 40 per cent or more of gross national product is provided by agriculture.

2 The percentage of GNP provided by agriculture is very low in Libya and Saudi Arabia. Why?

3 Study the map on this page and the map on page 22, and write a short account of 'The relationship between the relative wealth of a country and the percentage of GNP provided by agriculture.'

4 Does the relationship suggest **a)** that farming is unimportant in the Third World? **b)** that too much food is produced in the Third World? or **c)** that other sorts of production should make a greater contribution to the wealth of Third World countries?

The share of agriculture in GNP, by country

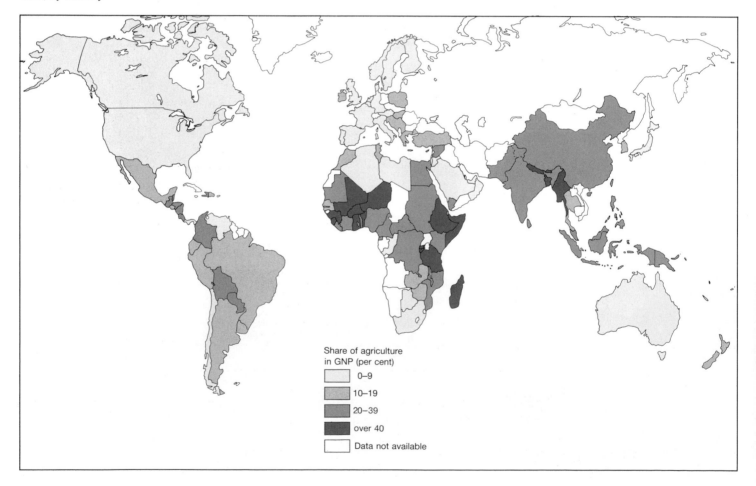

Share of agriculture
in GNP (per cent)

	0–9
	10–19
	20–39
	over 40
	Data not available

While some large-scale industrial developments have proved valuable, many people now argue that in some countries it would be better to spend scarce resources on smaller-scale and more widespread industrial activity.

These exercises look at Zambia in Southern Africa. Zambia achieved independence in 1964, but previously had been a British colony known as Northern Rhodesia. In 1980 about 73 per cent of Zambia's workforce was involved in farming, although the country imports much of its food, especially in years of drought. Mineral exports account for the bulk of its income. Industries other than metal ore smelting are of only local significance. Much of the power for industry is

provided by hydro-electric power stations in the Kariba and Kafue river gorges.

3.9 Industrialisation

1 Describe and account for the different manufacturing activities of Mbala, Kitwe, Lusaka (the capital) and Livingstone.
2 Compare Zambia, a Third World country, with the United Kingdom, under the headings of **a)** population features, **b)** the jobs people do, and **c)** the wealth of the country.
3 Write a few sentences about manufacturing industry in Third World countries.

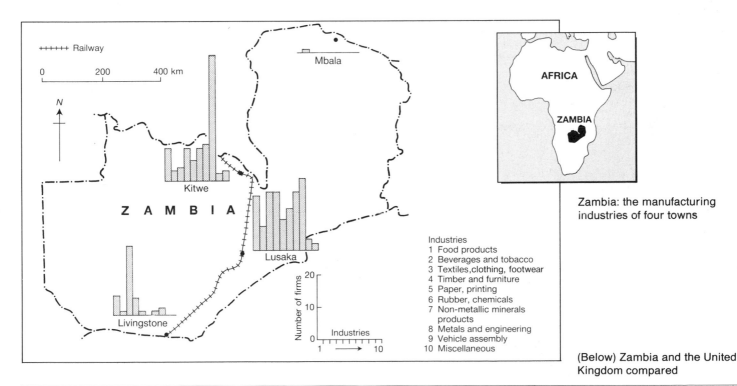

Zambia: the manufacturing industries of four towns

(Below) Zambia and the United Kingdom compared

	Area (km²)	Population millions 1985	GNP per capita US $ 1985	Population features			Labour force 1980		
				Births per thousand 1985	% aged 0–14 1980	% Urban 1985	% in Agriculture	% in Industry	% in Services
Zambia	752 614	6.7	390	49	46.9	48	73	10	17
United Kingdom	244 046	56.5	8460	13	21.1	92	3	38	59

3.10 Urbanisation and the urban economy

POPULATION IN MILLIONS	1950		2000
New York – NE New Jersey	12.3	Mexico City	31.0
London	10.4	Sao Paulo	25.8
Rhine-Ruhr	6.9	Shanghai	23.7
Tokyo – Yokohama	6.7	Tokyo – Yokohama	23.7
Shanghai	5.8	New York – NE New Jersey	22.4
Paris	5.5	Beijing (Peking)	20.9
Buenos Aires	5.3	Rio de Janeiro	19.0
Chicago – NW Indiana	4.9	Bombay	16.8
Moscow	4.8	Calcutta	16.4
Calcutta	4.6	Jakarta	15.7
Los Angeles – Long Beach	4.0	Los Angeles – Long Beach	13.9
Osaka-Kobe	3.8	Seoul	13.7
Milan	3.6	Cairo	12.9
Bombay	3.0	Madras	12.7
Mexico City	3.0	Buenos Aires	12.1

Both the 1950 and the 2000 population figures refer to the 'urban agglomeration' (usually the same as the metropolitan area population). Projections for the year 2000 are based on past trends and the country's economy, population growth and population movement.

A major feature of recent decades has been the rapid growth of huge urban areas. This has been partly due to high birth rates, but also to massive migration to the cities from the countryside. This rapid growth has led to a 'dual economy' in most Third World cities.

1 Put into words what the diagram shows about cities with populations of over 1 million. With the help of the table and a world map, try to identify the five cities with populations of over 10 million in the early 1980s.

2 Which of the cities expected to have over 12 million people by the year 2000 were not in the 'Top Ten' in 1950? Which of these are in the North, and which in the South?

3 In what ways is rapid growth of huge cities a help or hindrance to the economic and social development of countries in the Third World?

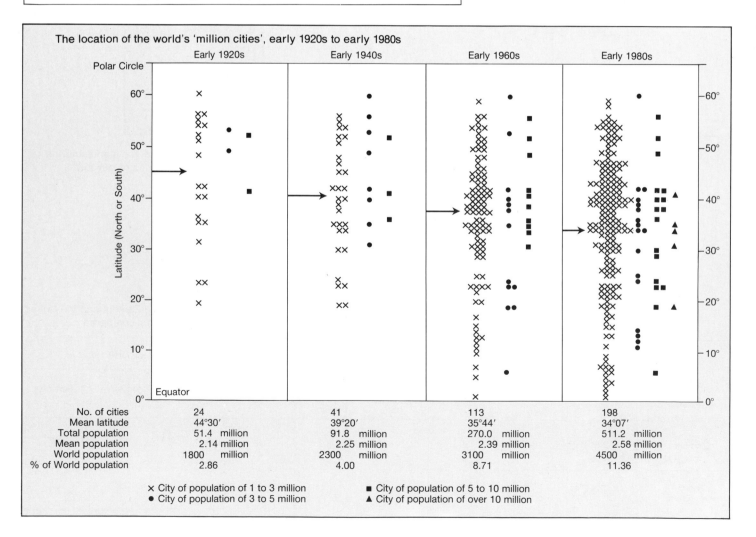

The location of the world's 'million cities', early 1920s to early 1980s

	Early 1920s	Early 1940s	Early 1960s	Early 1980s
No. of cities	24	41	113	198
Mean latitude	44°30′	39°20′	35°44′	34°07′
Total population	51.4 million	91.8 million	270.0 million	511.2 million
Mean population	2.14 million	2.25 million	2.39 million	2.58 million
World population	1800 million	2300 million	3100 million	4500 million
% of World population	2.86	4.00	8.71	11.36

× City of population of 1 to 3 million   ■ City of population of 5 to 10 million
● City of population of 3 to 5 million   ▲ City of population of over 10 million

The role of women varies greatly around the world, but almost everywhere their contributions to the economy and the work they do is unequally rewarded when compared with that of men. They do not have the same opportunities, and often have less access to education, to employment in responsible jobs, and to power in society as a whole. The degree of inequality and disadvantage varies from country to country.

The table looks at women in Muslim countries.

1 Draw statistical diagrams to show two Muslim countries where women seem most disadvantaged and two where they appear more equal with men *according to the four indicators used in the table*.

2 Consider each of the indicators used in the table and say whether you think it is a good or bad measure of the equality of women with men. What other indicators might be used?

3 Can a country be said to be developed if women experience disadvantages? What evidence is there of equality and/or disadvantage for women in Britain?

Women in 28 Muslim countries, 1980						
	Percentage of women aged 15–19 married	Percentage of adults illiterate		Percentage enrolment in school 12–17 yrs		Percentage in total labour force	
		Female	Male	Female	Male	Female	Male
Northern Africa							
Algeria	44	71	44	35	57	2	43
Egypt	30	59	33	30	54	5	51
Libya	72	61	25	22	64	3	46
Morocco	31	85	63	25	44	8	44
Sudan	43	94	64	17	32	7	55
Tunisia	6	64	37	25	45	4	44
Western Africa							
Gambia	–	97	73	11	23	43	55
Guinea	52	64	46	40	59	30	44
Mali	79	98	81	10	22	50	57
Mauritania	66	–	–	9	29	3	59
Niger	81	99	90	6	11	6	57
Senegal	63	94	67	25	38	31	52
Eastern Africa							
Somalia	14	99	90	22	42	22	55
Middle Africa							
Chad	73	99	64	6	20	18	60
South–West Asia							
Iraq	32	77	37	36	72	2	46
Jordan	28	53	31	54	70	3	24
Kuwait	28	46	31	59	67	5	46
Lebanon	13	32	15	58	72	10	42
Saudi Arabia	–	98	70	30	43	3	49
Syria	28	62	29	47	80	6	44
Turkey	21	51	19	36	58	31	52
Yemen, N	50	99	84	3	25	2	52
Yemen, S	–	89	33	27	66	3	48
Middle South Asia							
Afghanistan	49	99	78	5	26	13	53
Bangladesh	72	82	37	11	37	12	54
Iran	45	72	35	48	76	8	47
Pakistan	31	91	46	7	21	6	47
South-East Asia							
Indonesia	32	42	23	34	46	20	48
United Kingdom	11	20	6	90	79	32	60

Data not available for the following Muslim countries:
Western Sahara, Comoros, Djibouti, Bahrain, Oman, Qatar, United Arab Emirates, Maldives and Albania.

3.12 Conflicts within countries

In some parts of the world the great efforts made towards economic and social development are undermined by the enormous waste of human life and resources on internal conflicts.

The map shows conflicts that were taking place at the beginning of 1987.

1   What is the difference between international, civil and secessionist wars?

2   Study the map. **a)** Which of these wars have now ceased? **b)** Have any wars been started or renewed since January 1987? **c)** Name and locate other wars that you think might have been shown on the map.

3   Why are wars of any sort a constraint on economic development for the country or countries involved? Why might it be an economic help for other countries? Why is it sometimes claimed that war leads to longer-term development?

Conflicts around the world, 1987. Some of these have been going on for many years

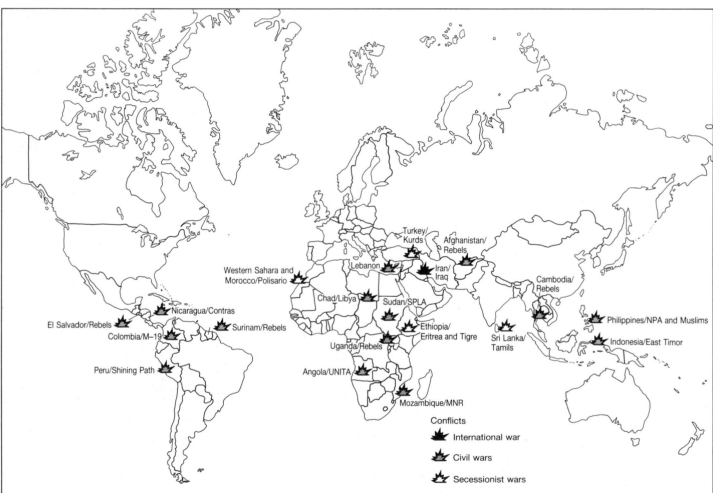

Western Sahara and Morocco/Polisario

Nicaragua/Contras

El Salvador/Rebels

Colombia/M–19

Surinam/Rebels

Peru/Shining Path

Chad/Libya

Lebanon

Turkey/Kurds

Afghanistan/Rebels

Iran/Iraq

Sudan/SPLA

Ethiopia/Eritrea and Tigre

Uganda/Rebels

Angola/UNITA

Mozambique/MNR

Cambodia/Rebels

Sri Lanka/Tamils

Philippines/NPA and Muslims

Indonesia/East Timor

Conflicts

International war

Civil wars

Secessionist wars

3.13 South Africa and apartheid

A particularly extreme form of inequality is that stemming from the policy of separate development or apartheid in the Republic of South Africa. Through this the hopes and aspirations of many people are frustrated by the enforced wishes of others with greater power.

South Africa is one of the richest and most economically developed countries in Africa, but it is also going through a period of intense and often violent change. The population is very mixed, and living standards and quality of life varies within and between the various groups. The Blacks have no political power or rights outside the homelands.

1 Draw diagrams to show the proportions of the various population groups in a) the whole country and b) each province. What important fact do these diagrams illustrate?

2 Look at the population figures. The various population groups are not spread evenly among the provinces. In which provinces are a) the Coloureds and b) the Indians concentrated?

3 On what grounds are equal political and social rights not available to everyone in the provinces?

4 Describe the features of the homelands shown by the map. What proportion of Black South Africans live in them? Why do the majority of Black South Africans not see 'separate development' in the homelands as the best means of bringing about Black advancement or development?

South Africa and the homelands, and population statistics

TOTAL POPULATION AT MID–1983 31 305 607		
Group	Group total	% of total population
Black	22 803 688	72.84
White	4 822 745	15.40
Coloured	2 803 174	8.95
Indian	876 000	2.80

WHITES

Province	Population mix	Number	% of total White population	% of Province population
Transvaal	Various	2 586 441	53.63	28.90
Cape	Various	1 289 080	26.73	23.99
Natal	English	586 018	12.15	20.62
Orange Free State	Mainly Afrikanar	334 998	6.94	16.10
Various homelands	Various	26 208	0.55	0.22

INDIANS

Province	Number	% of total Indian population	% of Province population
Natal	706 691	80.67	24.87
Transvaal	127 423	14.55	1.42
Cape	33 204	3.79	0.62
Orange Free State	None	None	None
Various homelands	8 682	0.99	0.07

COLOUREDS

Province	Number	% of total Coloureds population	% of Province population
Cape	2 357 990	84.12	43.88
Transvaal	266 268	9.50	2.97
Natal	95 479	3.41	3.36
Orange Free State	59 635	2.12	2.87
Various homlands	23 802	0.85	0.20

BLACKS

Homeland	Tribe	Number	% of total Black population
KwaZulu	Zulu	3 778 602	16.57
Gazankulu	Tsonga	582 508	2.55
Lebowa	North Sotho	1 862 514	8.17
Qwa Qwa	South Sotho	305 000	1.34
KwaNdebele	South Ndebele	225 792	0.99
KaNgwane	Swazi	183 763	0.81
Transkei	Xhosa	2 524 353	11.07
Ciskei	Xhosa	728 441	3.19
Bophuthatswana	Tswana	1 433 424	6.29
Venda	Venda	376 470	1.65
Total Black homeland population		12 000 867	52.63
% of total homeland population 99.51			

Non-homeland population

Province	Tribal mix	Number	% of total Black population	% of Province population
Transvaal	Various	5 970 354	26.18	66.70
Cape	Mainly Xhosa	1 693 507	7.43	31.51
Orange Free State	Various	1 685 427	7.39	81.03
Natal	Mainly Zulu	1 453 533	6.37	51.15
Total Black non-homeland population		10 802 821	47.37	
% of total non-homeland population 56.13				

3.14 Refugees from conflict

One of the worst features of the modern world is the existence of millions of refugees who have fled from or been forced to leave their country. They often survive mainly through the generosity and help of other people.

The table shows the top twenty countries according to the ratio of refugees to local population; other countries not appearing in the top twenty might also have many refugees, such as the UK and USA.

1 Choose three countries from the list and suggest where the majority of refugees came from and the reasons why they left their original country.

2 Identify some of the problems often faced by refugees in leaving their country, in making the journey to a new country, and upon arrival in the new country.

3 Which of the receiving countries are in the South? Why is coping with refugees a great problem in these countries?

4 Which of the world's wealthiest countries are in the 'Top Twenty'? In what ways might some countries be contributing to the support of refugees although they are not in the table?

5 In 1985 the majority of refugees entering the UK came from Iran and Sri Lanka. What were the circumstances in those countries that caused these people to leave?

Ratio of refugees to local population: top 20 countries			
	Population (millions) 1985	Refugees 1986–87	Ratio of refugees to local population	GNP per capita US$ 1985
Somalia	5.4	700 000	1 in 8	280
Iran	44.6	2 600 000	1 in 17	—
Burundi	4.7	267 500	1 in 18	230
Belize	0.2	9 000	1 in 22	1190
Sudan	21.9	974 000	1 in 22	300
Djibouti	0.4	16 700	1 in 24	—
Pakistan	96.2	2 882 000	1 in 33	380
Zambia	6.7	138 300	1 in 48	390
Honduras	4.4	68 000	1 in 65	720
Swaziland	0.8	12 100	1 in 66	670
Malawi	7.0	100 000	1 in 70	170
Sweden	8.4	120 000	1 in 70	11 890
Canada	25.4	353 000	1 in 72	13 680
Costa Rica	2.6	31 200	1 in 83	1 300
Angola	8.8	92 200	1 in 95	—
Tanzania	22.2	220 300	1 in 101	290
Uganda	14.7	144 000	1 in 102	—
Zaire	30.6	301 000	1 in 102	170
Yemen, Arab Rep.	8.0	75 000	1 in 107	550
Lesotho	1.5	11 500	1 in 130	470
UK	56.5	100 000	1 in 565	8 460

The ratio of refugees to local population is often higher than indicated because the refugees are concentrated in particular areas.

UK Refugees			
Place of origin	Given asylum	Leave to stay	Total
Iran	357	666	1023
Sri Lanka	907	33	940
Iraq	95	161	259
Uganda	44	89	133
Ghana	75	45	120
Ethiopia	41	16	56
Poland	23	21	44
Rest of world	168	128	296
Total	836	2032	2868

1985 figures

Many countries that were colonies and parts of empires are now independent states. Some show certain gains from those times, but others still suffer from their colonial experiences and the present day consequences of their colonial past.

These exercises look specifically at the countries of South America.

1 Choose a suitable key and on an outline map showing the countries of South America shade in the information about official languages given in the table.

2 How does the position of the 'Pope's Line' help explain the pattern of languages shown on your map?

3 Which country is named after a) its location, b) a mineral resource, c) a colonising country in Europe, d) a leader who led the country to independence?

4 With the help of this data, and any other evidence, write a short account of 'The colonies and empires of South America'.

South America

Country	Official language	Dominant religion
Argentina	Spanish	Roman Catholicism
Bolivia	Spanish, Quechua, Aymará	Roman Catholicism
Brazil	Portuguese	Roman Catholicism
Chile	Spanish	Roman Catholicism
Colombia	Spanish	Roman Catholicism
Ecuador	Spanish	Roman Catholicism
French Guiana	French	Roman Catholicism
Guyana	English	Various Christian
Paraguay	Spanish	Roman Catholicism
Peru	Spanish	Roman Catholicism
Surinam	Dutch	Christianity, Hinduism, Islam
Uruguay	Spanish	Roman Catholicism
Venezuela	Spanish	Roman Catholicism

4.2 The European 'Scramble for Africa'

One of the clearest examples of colonialism is the way Europeans took control of large areas of Africa by force during the last century.

These three cartoons present a critical view of the motives and methods of the European colonisation of Africa.

1 Look at the cartoons. What point is being made by the cartoonist in each case?

2 What *were* the motives of Europeans in colonising Africa? How did they benefit?

3 What are some of the problems faced by African countries that might be said to be rooted in their colonial past?

Cartoon views of German (right), British (below left) and Belgian (below right) behaviour in Africa

4.3 Exploitation and independence

European nations and companies often gained a great deal from the exploitation of resources and people around the world. Most colonies have now gained their independence, but are still affected by their past status.

The extract examines the economic outlook for St Kitts-Nevis at the time of its independence in 1983. St Kitts has since reverted to the name St Christopher.

1 Describe the new country's location and shape.

2 What are seen as the country's main economic strengths, and its main economic needs?

3 What does the extract say about the colonial past of St Kitts-Nevis?

4 What is likely to be some of the evidence on the islands of their colonial past?

5 In what ways, if any, should a colonising nation help its old colonies when they have gained independence? Why should they do what you suggest?

The photograph (below left) shows sugar cane being harvested

Independence will leave economy vulnerable
Agriculture and tourism the keys to survival

When the twin-island Caribbean state of St Kitts-Nevis becomes independent of Britain with the blessings of Princess Margaret and a golden handshake from the British Government, its biggest problems are going to be economic, not political.

The Prime Minister, Dr Kennedy Simmonds, will face a common problem – how to engineer development without adequate capital, and how to attract the sort of investment that will benefit the islands at least as much as the investors.

The historic backbone of the islands' economy has been sugar. In 1980, three and a half centuries after the British first settled on St Kitts, the crop still accounted for more than half the cultivated land, a quarter of all jobs, 20 per cent of gross domestic product and about two-thirds of exports. But with international sugar prices miserably low, the industry is losing money fast – about £4.5m in 1981.

Understandably, the aim of the Government is to diversify into new crops and light manufacturing. 'Agriculture must be the backbone of any economic restructuring,' Dr Simmonds says. 'We want to see agriculture providing the raw materials for the establishment of agro-industries.'

As in most of the Caribbean, seducing young people back to the land is a formidable task for a young state. St Kitts already has a small light manufacturing sector, comprising mainly electronics, clothing, shoes, soft drinks and beer, and batik products. But new investment is hard to come by.

The other sector with potential for expansion is tourism. About 40 000 visitors a year arrive in the deep water harbour off Basseterre, the capital, or the newly refurbished Golden Rock airport outside town. But in the recession of recent years St Kitts-Nevis has suffered by not being internationally known as a tourist destination and through limited direct access.

While the islands grope for a substitute for sugar, and with imports easily outweighing exports, the basic economic picture is depressingly similar to those of many of their eastern Caribbean neighbours.

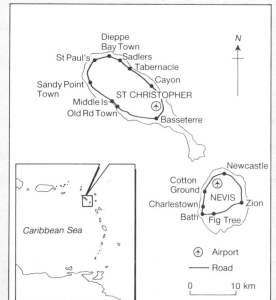

Dieppe Bay Town
St Paul's — Sadlers
Tabernacle
Cayon
Sandy Point Town
ST CHRISTOPHER
Middle Is
Old Rd Town
Basseterre
N

Caribbean Sea

Newcastle
Cotton Ground
Charlestown — NEVIS — Zion
Bath — Fig Tree

✈ Airport
— Road
0   10 km

4.4 World trade

Many Third World countries are involved in world trade and the global economy. Generally speaking they are the weaker partners in this trade, and are very dependent on changing demands from other places, and on decisions made elsewhere.

Minerals, especially copper and cobalt, provide over 95 per cent of Zambia's export earnings. The future development of the country, both economically and socially, depends on these earnings.

The Zambian copper belt and the railway network of central Africa

- /////// Copper belt
- ——— Line in use 1954
- ·········· Line opened 1955–80
- ═══ Water route

1 What happened to the price of copper between 1970 and 1982? What happened to the real value of income from copper during that time? What did that mean, in simple terms, to the income of the country, even if people worked as hard and as efficiently throughout that period?

2 What happened to the cost of imports during the same period? What were many imports likely to have been? Why did the prices of imports change during that period?

3 Most countries are dependent on world trade for their wealth and well-being. Why are so many Third World countries badly affected, and their plans for development greatly hindered, by conditions of world trade?

4 What non-financial handicaps does Zambia face in engaging in world trade?

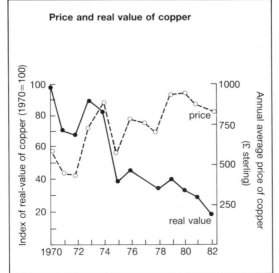

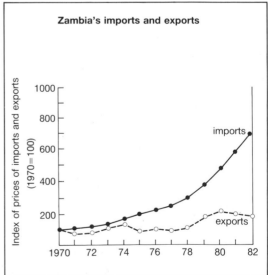

226

4.5 Transporting Third World commodities

The physical movement of goods remains a vital part of world trade. Many Third World countries are handicapped by having inadequate internal transport systems, or air and seaports that cannot cope properly with international trade.

The extract looks at the opening of a railway line in Gabon.

1 Describe the location and features of Gabon from the evidence provided.

2 Describe some of the features of the railway line opened in December 1986.

3 What raw materials are likely to be carried by this line? What arguments were given for and against the original proposal, and how was the project financed? Why might the European firms have become involved in such a project?

4 Why do you think the French Prime Minister, Monsieur Jacques Chirac, was invited to open the railway?

The Trans-Gabon railway line was built across some very difficult terrain

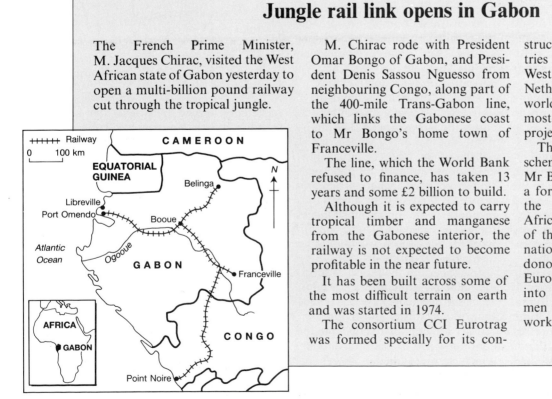

Jungle rail link opens in Gabon

The French Prime Minister, M. Jacques Chirac, visited the West African state of Gabon yesterday to open a multi-billion pound railway cut through the tropical jungle.

M. Chirac rode with President Omar Bongo of Gabon, and President Denis Sassou Nguesso from neighbouring Congo, along part of the 400-mile Trans-Gabon line, which links the Gabonese coast to Mr Bongo's home town of Franceville.

The line, which the World Bank refused to finance, has taken 13 years and some £2 billion to build.

Although it is expected to carry tropical timber and manganese from the Gabonese interior, the railway is not expected to become profitable in the near future.

It has been built across some of the most difficult terrain on earth and was started in 1974.

The consortium CCI Eurotrag was formed specially for its construction by 19 firms from six countries – Belgium, Britain, France, West Germany, Italy and the Netherlands. It is one of the world's, as well as black Africa's, most ambitious civil engineering projects.

The World Bank considered the scheme economically unsound. But Mr Bongo, whose oil-rich country, a former French colony, is one of the smallest but wealthiest in Africa, argued that it was the 'spine' of the economy and a symbol of national unity, and other aid donors, led by France and the European Community, stepped into the breach. More than 4 000 men from about 20 countries worked on it.

4.6 Multinational corporations

Third World countries are increasingly affected by the activities of multinational and global corporations. While this can result in employment and growth, it does make the well-being of many people dependent on decisions made elsewhere. The interests of shareholders are likely to be of more importance than the needs of the people and country where the branch or subsidary is located.

The extract looks at the multinational mining companies in Namibia. South Africa has occupied Namibia since 1915, although this was declared illegal by the International Court of Justice in 1971; SWAPO, the South West Africa People's Organisation, is the movement fighting for independence.

1 Why are multinational companies operating in Namibia being criticised by the United Nations, the International Court of Justice and the Catholic Institute for International Relations?

2 What arguments are presented by the mining companies for their presence in Namibia? What is their attitude to the possibility of an independent Namibia in the future? What changes have some companies recently introduced to their operations?

3 What are the advantages and disadvantages of such economic activity for **a)** the multinational companies and **b)** the people of the host country?

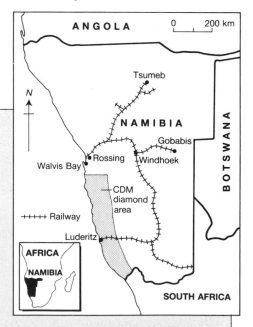

Namibia

In the shadow of the mining giants

British mining giants like the Rio Tinto Zinc Corporation come under heavy attack for their role in supporting South Africa's illegal occupation of Namibia, in a report by the independent Catholic Institute for International Relations (CIIR).

Although independent pressure from the outside, coupled with an attempt by these multinationals to improve their own public and employee relations at home in anticipation of an independent Namibia, has led to some improvement in the living and working conditions of many of their black employees, the blacks are still far from receiving equal status with their white counterparts, the report points out.

The CIIR says that if anybody benefits from the occupation of Namibia it is these international mining companies and South Africa. It shares the view expressed by Swapo, that Namibians derive little or no benefit from their operations.

An independent Namibia, however, will heavily depend on continued operations of the existing mines. They will bring in the much-needed foreign capital.

Drastic changes will have to be wrought in the management structure of these outfits. At present these skills are almost nonexistent among the Black Namibians, mainly because of the lack of education and training facilities.

Some of the multinationals have only recently begun a training and educational programme. The report points out that many of the workers believe that 'the training programme is a joke or a public relations exercise and what particularly rankles is the company's refusal to promote a qualified man unless there is a vacancy which he can fill: workers see this as a way of holding down Black Namibians'.

In conclusion the report appreciates that on some mines considerable improvements have been made to workers' wages and living conditions, but goes on to condemn the continued use of contract labour, which prevents families from joining their men. Although Rossing in the last few years has begun to redress this situation with its new family townships.

Multinationals like RTZ have repeatedly argued that since consecutive British governments have refused to endorse sanctions, they do not see themselves as breaking any international laws.

The attitude therefore seems to be that the country would have more need for these multinationals to stay, rather than the companies being desperate to be allowed to stay.

This attitude was confirmed 12 months ago by Rossing's chairman Mr Ronnie Walker when he said that 'here in Namibia they would be frightful fools to throw us out – they only have to see our track record to recognize the company's importance to an independent Namibia.'

He has publicly said that his company would be prepared to work in close co-operation with any incoming government.

Third World countries are at a disadvantage in the global economic systems for a number of reasons. They often depend on one or two export products that are very vulnerable to world prices. They may need to import products that they cannot produce themselves and are again vulnerable to rapidly rising prices. Surplus produce from elsewhere may be sold at low prices in Third World countries, so making it difficult for local producers to compete.

Another problem is the 'dumping' of many dubious products on Third World markets.

4.7 Third World imports

1 Why is the dumping of pharmaceuticals (medicinal drugs) and pesticides in Third World countries a) relatively easy to do, and b) so harmful?
2 What other imports into Third World countries not mentioned here are a) dangerous to people, and b) a hindrance to economic development?
3 If the dumping of products is harmful to people and the economies of countries, why does it happen?

Dumping

Dumping is:

- Selling goods banned elsewhere.
  For example: Exporting TRIS-treated childrens' nightwear after TRIS was found to be carcinogenic in the US.
- Selling goods casually which are highly restricted and controlled elsewhere.
  For example: Prescription-only drugs sold on marketplace stalls in Indonesia.
- Selling goods in an environment for which they haven't been designed and in conditions which can make them unsafe.
  For example: Babyfoods promoted where water to dilute the powder is dirty, where illiterate parents cannot read the instructions and families are too poor to afford enough formula for regular and adequate feeds.
- Transferring polluting industries and/or unsafe manufacturing processes to countries where unions are weak and governments acquiescent.
  For example: Asbestos manufacture in South-East Asia.

This billboard in Bombay highlights the dangerous aspects of the drugs industry in India, some of which might be attributed to dumping

Chemical products

Pesticides, drugs and industrial chemicals are all chemical products. Because of their toxicity, they are amongst the most dangerous products that can be offloaded onto unprotected consumers.

The dumping of pesticides in the Third World fields can occur because:
- Lack of restrictions means pesticides can be sold even though they may be banned elsewhere.
- The users are likely to be rural workers with an unjustified faith in Western technology, and perhaps unable to read warning and mixing instructions.
- There are often few or no label warnings.
- Little or no protective clothing, masks or gloves are worn.

The temptation to dump chemical products in the Third World is high because:
- There are few or no controls over testing or registration of imports.
- There are few or no obligations to print appropriate warnings or advice on the labels.
- There are few or no restrictions of supply, which might ensure responsible control through safe outlets.

The risks are greater because those working with the chemicals in the Third World are less aware of the dangerous consequences or spilling, breathing, handling or ingesting the products.

The dumping of pharmaceuticals in the Third World can occur because:
- Dangerous drugs are available without prescription from untrained pharmacy staff.
- Drug labels, package inserts and adverts to doctors sometimes give no warnings on side-effects, optimal dosages vary.
- Often there are no shelf-life restrictions.
- Little or no testing is done on drug imports.
- Drugs withdrawn in the West are often available.
- Most causes of death go unrecorded. Sources of illness are difficult to find. So drug-induced problems are impossible to trace.

4.8 Third World loans and debts

Sudan ensnared in classic Third World debt trap

A collapsed economy and seething social unrest led the Prime Minister, Sadiq al-Mahdi, to introduce last weekend a one-year state of emergency. Sudan's fledgling democracy is increasingly under pressure, as popular expectations founder on the government's inability to do anything about the key ills afflicting the nation – the civil war in the south, and the foreign debt.

Sudan's total external debt lies between $10bn and $13bn. Just servicing it would cost the country much more than what it earns from exports. Sudan is caught in the classic dilemma of Third World debtors. Its decline has gone so far that it can no longer realistically help itself.

Argentina to sign $30bn debt refinancing

A top-level Argentine delegation arrives in New York today to sign an accord refinancing more than $30bn of the country's outstanding debt and setting the seal on $1.95bn of new money.

Under the terms of the agreement payments on the debt principal will be pushed back until 1992 and after. Until then, Argentina will make only interest payments.

The new money is to be provided by commercial banks and marks the first such package put together under the so-called menu approach to the debt problem.

The system was designed to overcome banks' growing reluctance to make fresh loans to indebted nations.

Peruvian bank plan

President Garcia's announcement on Monday night that he planned to nationalise Peru's banks, including six foreign institutions, has been greeted with hostility and scepticism in the international banking world. The move will increase the isolation of Peru, already shunned by foreign bankers because of Garcia's policy of limiting service payments on its $14bn debt to 10 per cent of export revenues.

One of the greatest problems facing most of the poorer nations of the world is their huge debt. They borrowed from governments and commercial banks for what they needed, but as the interest rates have changed many are unable to repay what they owe. In many cases annual debt repayments are as much as total earnings.

1 Draw a statistical diagram showing the debt-service payment percentages given in the table for the twelve countries.

2 Why are commercial American and British banks more concerned about debts owned by some countries than others? Suggest a few examples of each type of country.

3 Some Third World countries argue that they are unfairly trapped because they borrowed when interest rates were low, and now they have to pay unexpectedly high interest rates. Do you think they have any cause to complain? What should they do to repay these debts – most are already very poor?

4 The extracts look at some of the events in the summer of 1987. Explain the headline about Sudan. What were Argentina and Peru doing to try to overcome their debt problem?

5 What sorts of conditions do the World Bank and International Monetary Fund usually lay down before giving a loan to a country?

Developing countries – the debts in 1985

Debt to American banks	$ billions	Debt to British banks	$ billions
Mexico	25.8	Brazil	9.3
Brazil	24.8	Mexico	8.7
South Korea	11.1	South Africa	5.6
Venezuela	10.4	Argentina	3.4
Argentina	8.4	South Korea	2.7
Chile	6.4	Venezuela	2.7
Philippines	5.1	Nigeria	2.4
South Africa	3.9	Chile	2.1
Taiwan	3.1	Philippines	1.7
Indonesia	3.1	Malaysia	1.7

Developing countries debt service payments

The first figure indicates what each of the countries below owed as of 1986; the second indicates their debt service ratio, i.e. the payment of debt and interest as a percentage of exports of goods and services.

Country	Debt ($US billions 1986)	Ratio (percentage 1986)
Brazil	97	42
Mexico	91	52
Argentina	43	64
South Korea	34	24
Venezuela	32	37
Indonesia	36	33
Philippines	22	21
Turkey	24	32
Algeria	15	55
Nigeria	22	23
Chile	18	37
Peru	12	21

So Mexico had to repay more than half the foreign exchange it earnt in 1986, to overseas lenders. Brazil had to repay 42 per cent of all its export earnings that year.

Many governments give a small part of their income to Third World countries as development aid. This is often dependent on business or construction contracts being given to the donor country. A great deal of aid that is used for smaller-scale local projects is provided by voluntary organisations from the donations and gifts of many individuals.

1 Graph the data on official development assistance from OECD countries as a percentage of donor GNP in 1985. Which countries exceeded the target of 0.7 per cent of GNP suggested by the UN?

2 Describe and explain the location and type of countries receiving most of Britain's bilateral aid in 1985.

3 Describe the main details of the Malaysian water supply project that was helped by aid from the British government.

4 What was the nature of the aid given by the government? What will Britain get out of the development project because it has given the aid? What other source of money was available?

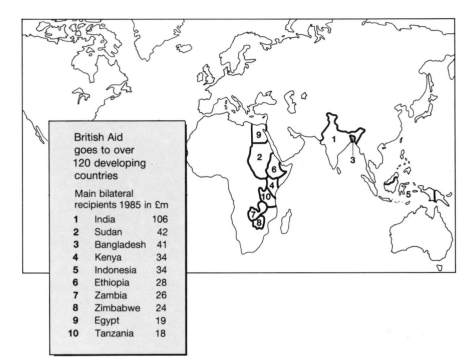

British Aid goes to over 120 developing countries

Main bilateral recipients 1985 in £m

1	India	106
2	Sudan	42
3	Bangladesh	41
4	Kenya	34
5	Indonesia	34
6	Ethiopia	28
7	Zambia	26
8	Zimbabwe	24
9	Egypt	19
10	Tanzania	18

Official development assistance as a percentage of donor GNP (OECD countries)		
	1975	1985
Italy	0.11	0.31
New Zealand	0.52	0.25
United Kingdom	0.39	0.33
Belgium	0.59	0.53
Austria	0.21	0.38
Netherlands	0.75	0.90
France	0.62	0.79
Japan	0.23	0.29
Finland	0.18	0.39
West Germany	0.40	0.48
Denmark	0.58	0.80
Australia	0.65	0.49
Sweden	0.82	0.86
Canada	0.54	0.49
Norway	0.66	1.00
USA	0.27	0.24
Switzerland	0.19	0.31

Malaysian Water Project: British company and British aid

The Government may be forced to increase this year's budget for the Aid and Trade Provision (ATP) after final agreement of the financing package for a £410 million contract won by a Surrey company to provide drinking water systems in rural Malaysia.

The contract, awarded to Biwater, of Dorking, a private company specializing in water systems, will be supported by a record ATP grant of nearly £60 million.

The Malaysian project, initiated by Biwater in 1984, involves 174 separate water distribution schemes, some up to 3000 miles apart, to bring treated piped water to more than two million rural Malaysians.

Goods and services worth nearly £200 million will be supplied from Britain, with about half the value being sub-contracted by Biwater.

The chairman of Biwater said that 300 jobs would be created in the North West and West Midlands.

Loan agreements arranged by the Bank of America, worth £135 million, and backed by the Export Credits Guarantee Department, were signed yesterday and these, with the ATP grant, will cover the British work.

The Malaysian part of the project will be carried out by Biwater's 50 per cent-owned associate, which is jointly owned with a local company, Antah Holdings.

The project, won against French and Japanese competition, will take five years to complete.

Mr White said: 'It is a huge undertaking, in most inhospitable conditions.'

4.10 Does aid work?

There is a great deal of evidence that much aid does not help the people and communities most in need. This may be due to wrong decisions being taken about what projects are needed, or to restrictions being put on what the aid can be used for, or sometimes because of corruption by influential people.

These extracts present differing views about aid.

1   Why does Mark Brown criticise government to government aid? What is the view of the President of the World Bank about the giving of official aid?

2   What is said to be the strength of voluntary aid, as opposed to aid provided by governments or international banks?

3   What is the message of the cartoons? Do they really explain or do they over-simplify the arguments?

Third World aid must not be cut

The President of the World Bank replies to the critics of aid, particularly on the matter of waste and extravagance by Third World governments: 'Everything is imperfect and development aid misses now and then. But I would say the overwhelming weight of evidence speaks in the other direction. Take a look at the completed projects that have occurred in World Bank lending. The bank does not finance or tackle the support of any developing country unless we are convinced it will produce at least an economic rate of return of 10 per cent. The average of completed projects has been 17 per cent in bank lending. In the last four years the rate of return for agricultural projects has been 22 to 27 per cent on average. It's a good return in anybody's language.'

Mark Malloch Brown suggests a new approach to famine relief

Let Africa nourish its own roots

The pattern of public support for relief operations is depressingly familiar. It starts high and falls away as the recipients fail to get better with a 'quick fix' and some new issue intervenes to catch our attention. What Ethiopia and 20 other African countries need is long, patient support while they rebuild their farms.

This is hardly likely to electrify would-be donors. Nor, on past record, are governments and intergovernmental organisations necessarily best placed to give leadership. It is their vast, over-ambitious schemes that have got Africa into much of its present trouble.

It is people, not governments, who are bearing the brunt of the famine and it is individual people in the West who are helping them. The best bridges between these two groups are the voluntary agencies. At their best they provide an open and cost-effective way of helping. At the same time they could build enlightened but realistic long-term support for Africa.

More important than money is the engagement of sympathy, which needs to be channelled into support for long-term development before it dissolves. One agency has borrowed a Vietnamese proverb to make the point: give a man a fish and you give him a meal; give a man a net and he will never be hungry again. The voluntary agencies need to go on the offensive, both here and in Africa, as the advocates of small scale agricultural, and indeed urban, projects that will let Africans put themselves back on their feet.

An exciting lesson of contemporary Africa is that, despite drought, when Africans want something enough they make it work. What is crucial is that it fits with local aspirations. What does least well is official Africa. Too many governments are so distant from their people that they were the last to wake up to the threat of famine. Yet they are the ones through which the Western governments channel their aid.

Voluntary agencies can give only a fraction of the funds provided by governments, but they have the asset of an ear closer to the ground. They can pick winners. In-

creasingly the big governmental agencies use their money to put in more foreign experts to prop up institutions that Africans are not particularly interested in. These people, and there are tens of thousands of them with their families, and expatriate housing allowances, swallow ever larger chunks of the assistance available to Africa.

Because of the tension and hostility between the superpowers they often interfere in the affairs of Third World countries, aiming to gain the support of their governments and to prevent their opponents gaining any advantages.

The maps show the distribution of military advisers from the superpowers and military bases. They cannot be completely accurate because of the nature of the topic. The map of bases shows that the USSR's military strength is concentrated, in the USSR itself and Eastern Europe, whereas the USA's military strength is more widespread, with a greater number of bases spread around the world.

1 Describe the spread of military advisers shown by the map. What are the benefits for the superpowers of sending military advisers abroad, and of training foreign military personnel? How do the countries receiving military advisers hope to gain?

2 Name Third World countries shown on the map which provide military bases for **a)** the United States, and **b)** the USSR? Why should they allow foreign bases on their territory? What benefits do they get from such arrangements?

3 What possibilities or opportunities can foreign bases provide for the superpowers? How can a military presence abroad also lead to difficulties?

4 Write a few sentences about USSR involvement in Afghanistan, and USA involvement in Central America. What are the similarities and differences between the two superpowers in terms of **a)** motives and **b)** activities?

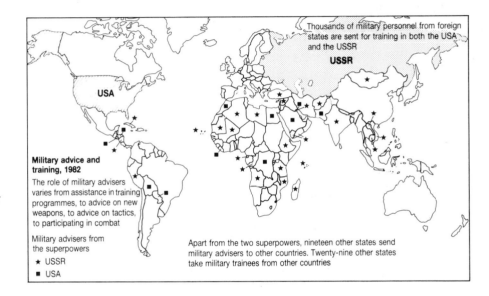

Thousands of military personnel from foreign states are sent for training in both the USA and the USSR

Military advice and training, 1982

The role of military advisers varies from assistance in training programmes, to advice on new weapons, to advice on tactics, to participating in combat

Military advisers from the superpowers
★ USSR
■ USA

Apart from the two superpowers, nineteen other states send military advisers to other countries. Twenty-nine other states take military trainees from other countries

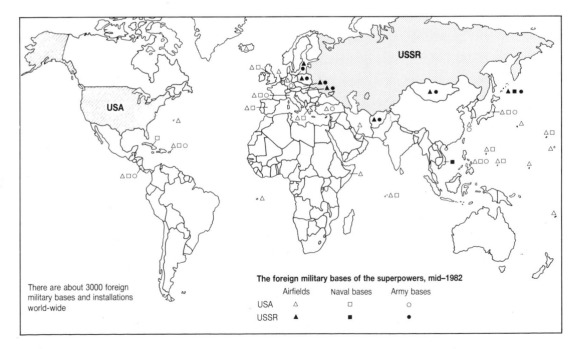

There are about 3000 foreign military bases and installations world-wide

The foreign military bases of the superpowers, mid–1982

	Airfields	Naval bases	Army bases
USA	△	□	○
USSR	▲	■	●

4.12 Development and dependency

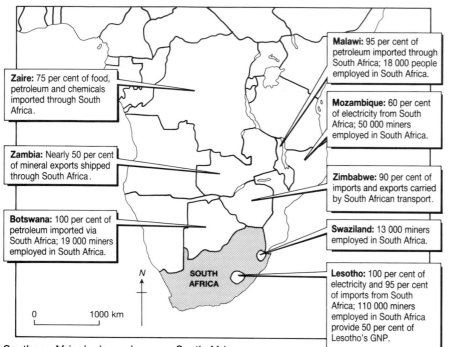

Zaire: 75 per cent of food, petroleum and chemicals imported through South Africa.

Malawi: 95 per cent of petroleum imported through South Africa; 18 000 people employed in South Africa.

Mozambique: 60 per cent of electricity from South Africa; 50 000 miners employed in South Africa.

Zambia: Nearly 50 per cent of mineral exports shipped through South Africa.

Zimbabwe: 90 per cent of imports and exports carried by South African transport.

Botswana: 100 per cent of petroleum imported via South Africa; 19 000 miners employed in South Africa.

Swaziland: 13 000 miners employed in South Africa.

SOUTH AFRICA

Lesotho: 100 per cent of electricity and 95 per cent of imports from South Africa; 110 000 miners employed in South Africa provide 50 per cent of Lesotho's GNP.

N

0   1000 km

Southern Africa's dependency on South Africa

The world is interdependent in economic, political and cultural ways. The development of Third World countries depends on the support of the more developed nations, as well as on the efforts of their own people.

The map, table and extracts look at aspects of dependency in Southern Africa.

1 Describe three distinct ways in which the economies of the states of Southern Africa depend on the actions or goodwill of the Republic of South Africa.

2 How might the black states of Southern Africa influence the economic and social development of the Republic of South Africa?

3 How do the economic, social and political relationships between the Republic of South Africa and the rest of the world illustrate global interdependence?

Southern Africa

It has been made clear that if third parties impose sanctions on South Africa, then South Africa will impose sanctions on its neighbours. If the neighbouring states themselves apply sanctions against South Africa, then South Africa might well retaliate with armed force. South African actions could cause already frail economies to collapse and fragile governments to fall.

As importers of South African goods and large users of South African transit facilities the black states can impose sanctions only at considerable cost to themselves. Exports vital for foreign exchange would be threatened and the inability to import foodstuffs could, in time of shortage, cause famine and starvation. South Africa, on the other hand, would only be slightly hurt by the loss of custom.

One of the few possibly effective actions African states could take against South Africa would be denial of air space for flights to and from Europe. This would hurt Whites rather than Blacks, would impede business contacts and hinder attempts to bust other sanctions.

Call to help frontline states

Foreign ministers of the 101 Non-Aligned Movement nations have recommended the establishment of a special 'solidarity fund' to help South Africa's black neighbours overcome the effects of sanctions.

The foreign ministers called on Third World nations to take the lead in supporting the threatened economies of the frontline states.

Leading world producers of selected metal ores			
	1	**2**	**3**
Chromium	South Africa	USSR	Albania
Gold	South Africa	USSR	Canada
Vanadium	South Africa	USSR	USA
Manganese	USSR	South Africa	Brazil
Platinum	USSR	South Africa	Canada
Asbestos	USSR	Canada	South Africa
Diamonds	USSR	Zaire	South Africa
Uranium	USA	Canada	South Africa

Index

This index is in two parts, a Place Index and a Subject Index. Page numbers in bold type identifiy the eight case studies which make up Part Two; page numbers in italic show where an entry appears in the Exercises (Part Three).

PLACE INDEX

SUBJECT INDEX

Oxford University Press, Walton Street, Oxford OX2 6DP

Oxford New York Toronto
Delhi Bombay Calcutta Madras Karachi
Petaling Jaya Singapore Hong Kong Tokyo
Nairobi Dar es Salaam Cape Town
Melbourne Auckland
and associated companies in
Berlin Ibadan

Oxford is a trade mark of Oxford University Press

© Oxford University Press 1989

ISBN 0 19 913329 8

Typeset by MS Filmsetting Limited, Frome, Somerset
Printed in Hong Kong